Engineering Speaking by Design

DELIVERING

TECHNICAL

PRESENTATIONS

WITH REAL

IMPACT

Engineering Speaking by Design

DELIVERING TECHNICAL PRESENTATIONS WITH REAL IMPACT

Edward J. Rothwell
Department of Electrical and Computer Engineering
Michigan State University, East Lansing, USA

Michael J. Cloud
Department of Electrical and Computer Engineering
Lawrence Technological University
Southfield, Michigan, USA

CRC Press
Taylor & Francis Group
Boca Raton London New York

CRC Press is an imprint of the
Taylor & Francis Group, an **informa** business

CRC Press
Taylor & Francis Group
6000 Broken Sound Parkway NW, Suite 300
Boca Raton, FL 33487-2742

© 2016 by Taylor & Francis Group, LLC
CRC Press is an imprint of Taylor & Francis Group, an Informa business

No claim to original U.S. Government works

Printed on acid-free paper
Version Date: 20150413

International Standard Book Number-13: 978-1-4987-0577-6 (Paperback)

Library of Congress Cataloging-in-Publication Data

Rothwell, Edward J.
    Engineering speaking by design : delivering technical presentations with real impact / author, Edward J. Rothwell and Michael J. Cloud.
        pages cm
    Includes bibliographical references and index.
    ISBN 978-1-4987-0577-6 (alk. paper)
    1. Communication of technical information. 2. Public speaking. 3. Communication in engineering. I. Cloud, Michael J. II. Title.

T10.5.R 63 2015
658.4'52--dc23   2015012023

Visit the Taylor & Francis Web site at
http://www.taylorandfrancis.com

and the CRC Press Web site at
http://www.crcpress.com

EJR: To Zach and Sam; "I have a son, who is my heart. A wonderful young man, daring and loving and strong and kind."
— Maya Angelou

MJC: To Leonid P. Lebedev, with gratitude

Contents

Preface

In *Life Without Principle*, Henry Thoreau laments hearing a speaker who chose "a theme too foreign to himself." As a result, there was "no truly central or centralizing thought in the lecture." A century and a half later, you may have encountered the engineering speaker who is lost in his own slides and resorts to reading them verbatim, or the engineer who, although capable of connecting with her audience, seems uninterested in doing so. In fact, that speaker may be you.

Engineers are smart people and their work is important. They make their accomplishments known through presentations to colleagues, bosses, customers, and the public. Their task, which is far more difficult than that of the typical public speaker, is to convey highly technical information in an often brief period of time. It is crucial that their formal presentations not be confusing, inaccurate, misleading, or boring. Heaped on that is the burden that the success of their entire endeavor may hinge on the first sentence uttered.

It shouldn't be surprising, then, that an engineer must make a concerted effort to become a skilled technical speaker. And yet in our experience as engineering educators we have found an attitude among students and colleagues alike that technical speaking is an *inherited gift*. ("Wow, Tom is such a natural speaker, I wish I had his talent.") What they don't realize is that this seemingly natural speaking style is hard-earned, paid for by hours of preparation, rehearsal, and self-examination.

In our previous book, *Engineering Writing by Design*, we made the case that engineers can become good writers by applying familiar concepts from engineering design. The same is true for technical speaking. We claim that *to speak like an engineer, you must think like an engineer*. Thus, you already have the tools to become a good formal speaker; you must only learn to apply them. That is our purpose in this book: to teach you to apply to your speaking tasks the design skills you have worked so hard to develop. You may find this key observation intuitively obvious, but our experience is that for most this viewpoint is a revelation.

A public speaking task can be seen and approached as a design problem accompanied by requirements, constraints, protocols, and standards to meet, and an eventual customer to satisfy (the target listener). Engineers are deeply familiar with design-oriented thinking, and their experience with design processes can be brought directly to bear on even the most daunting presentation tasks. This approach provides a familiar framework for the classical elements of public speaking skill that are taught in university courses and training seminars.

We have carefully organized this book to take you through the steps involved in the speaking process, from conception to delivery and beyond. Each chapter covers one step, ending with a checklist to help you keep on task. The early chapters deal with the steps of engineering, designing, building, and optimizing your presentation, and rely heavily on familiar principles of engineering design. The later chapters deal with delivery and aftermath. At each stage we encourage you to develop rubrics to gauge your progress; examples are given in Chapter 1. We feel that if you follow the process from beginning to end, you will become a confident and capable technical speaker. A checklist covering the entire process appears in the appendix.

And now a little on what this book is *not* about. The general topic of "engineering speaking" is obviously very broad; we choose to concentrate only on "formal" engineering presentations. These would include such things as in-class presentations, thesis or dissertation defenses, presentations at professional conferences, or presentations to potential customers or funding agencies. Our purpose is not to address one-on-one conversations, conference calls, conversations around the lunch table, meetings chaired under parliamentary procedure, etc. Such things are important, and certainly some of the principles we put forth will apply to spoken communication in *any* engineering setting. But for these, and for the many other types of speeches you may find yourself giving — welcome speech, keynote speech, nomination speech, acceptance speech, farewell speech, eulogy — we refer you to the many excellent general books on public speaking.

Acknowledgments

We would like to thank Jennifer Byford for her many insightful and useful comments on the manuscript, and for contributing multiple examples and exercises throughout the book. Beth Lannon-Cloud offered constructive criticism on several versions of the manuscript; she also generated a simple rubric for rating presentations. Andrew Szymczak offered helpful feedback and suggested an example for Chapter 1. We are indebted to Josh Myers and Junyan Tang for carefully reading the manuscript and providing helpful comments. Gordon P. Fellows suggested an example for Chapter 3.

We are grateful for the work of several anonymous reviewers. Our Taylor & Francis editors Nora Konopka, Michele Smith, and Laurie Oknowsky provided much valuable guidance and support throughout the publication process. The cover designer was John Gandour.

Authors

Edward J. Rothwell received the B.S. degree from Michigan Technological University, M.S. and E.E. from Stanford University, and Ph.D. from Michigan State University, all in electrical engineering. He has been a faculty member in the Department of Electrical and Computer Engineering at Michigan State University since 1985, and currently holds the rank of professor. Before coming to Michigan State he worked at Raytheon and Lincoln Laboratory. Dr. Rothwell has published numerous articles in professional journals involving his research in electromagnetics and related subjects. He is coauthor with Michael Cloud of *Electromagnetics* (CRC Press, 2nd ed., 2008) and *Engineering Writing by Design* (Taylor & Francis/CRC Press, 2014). Dr. Rothwell is a member of Phi Kappa Phi, Sigma Xi, the International Union for Radio Science (URSI) Commission B, and is a Fellow of the Institute of Electrical and Electronics Engineers.

Michael J. Cloud was awarded a B.S., M.S., and Ph.D. from Michigan State University, all in electrical engineering. He has been a faculty member in the Department of Electrical and Computer Engineering at Lawrence Technological University since 1987, and currently holds the rank of associate professor. Dr. Cloud has coauthored twelve other books, mostly in engineering mathematics. He is a senior member of the Institute of Electrical and Electronics Engineers.

1

Becoming a Presenter

1.1 Is Public Speaking Easy?

That's a tough question to answer. If like many people you find your knees buckling, hands sweating, and voice quivering each time you step in front of a large audience, you will say *No!* Others seem naturally comfortable when speaking in public. Every word seems to come out just right. To them it must be easy.

But what about *technical presentations*? Are they ever easy? While some people may have a knack for technical improvisation, we claim that *good technical presentations are never easy*. Many things must come together for success, including knowledge, clarity, confidence, integrity, and interest. Such a combination can only be expected from solid, intelligent preparation and much careful practice. If that sounds like a lot of work, consider the alternative.

Steve has always been the outgoing type. With a resonant voice, lively gestures, and friendly smile, he can entertain and even lead large groups of his friends in ordinary conversation. But recently Steve had to give an oral presentation in his junior level thermodynamics course. Feeling confident, he decided to "just be himself" and take a stab at the fifteen minute talk without notes, slides, or any real preparation. Five minutes into the talk, Steve realized he was struggling to make sense of his topic. He turned away from the dry erase board to face his audience for the first time:

> I realize I'm kinda all over the place here, but that's just how my mind works.

Everyone stared. Steve turned back to the board and tried to recover, but not even his biggest apologetic smile could save him. Embarrassed and a bit bewildered with himself, Steve ended his talk after just ten minutes. The instructor awarded a D grade rather than an F, but Steve knew the truth. His presentation was a failure.

Some preparation is better than none, but technical presentation is never routine. Care must be taken to make sure the preparation is both sufficient and appropriate for the speaking task at hand.

> Kelly, another thermodynamics student that semester, never dreamed of "winging" her presentation. On the contrary, she put in diligent work, carefully preparing slides and rehearsing her talk beforehand. But the longer Kelly spoke, the more she realized she really didn't understand her topic in any depth. She became increasingly nervous and unsure of herself, and it showed. Eventually she started reading her slides verbatim. Despite her best intentions, without sufficient preparation her presentation collapsed. Some listeners sensed her growing panic and looked away in embarrassment, while others became bored or even amused. In the end, Kelly's lack of confidence destroyed the effectiveness of her message.

For most of us, technical presentation is akin to musical performance: it is a complicated, learned skill. Talent helps but is not enough. Still, even a young performer, or technical presenter, can become proficient if he sets his mind to it. For engineers there is no option; the skill of technical speaking *must* be mastered. Regardless of your organization, your discipline, or the level of your position, you *will* find yourself delivering important talks to important people. When it's your turn to get up in front of that audience, you want to deliver something helpful, motivational, even inspirational. You want to earn the audience's respect and appreciation. In short, you want your technical talks to be good.

1.2   How Can I Learn to Be a Good Presenter?

Here's a basic roadmap. First, you'll need to understand the fundamentals of technical presentation. A general public speaking class will cover only some of these (and not every engineer has time to take a public speaking class anyway). So we'll provide these fundamentals in later chapters. Second, you need to do some homework by watching skilled presenters and noting how they implement the fundamentals. You'll probably spot some more sophisticated speaking techniques that go beyond the fundamentals; take particular note of these. Third, *practice makes perfect*. You'll need to practice speaking until the fundamentals become second nature. Private rehearsal is a good way to get started if you are inclined to be nervous; even gifted speakers rehearse in private before the big event. Truly useful experience will come in front of your colleagues or fellow students. Fourth, you'll have to stick with it. Growth comes through repeated practice. Sure, things may be rough at the start, but you will learn with each attempt. It won't be long before you get a taste of success. And finally, you'll have to keep your skills sharp by staying in practice. Don't forget this last point — it's easy to get rusty if you don't stay active. Seek out speaking opportunities during lulls. You won't regret it!

Regardless of how you get started, it's important to realize the serious and relatively complex nature of the technical speaking task. For Steve, the student subject of our recent story, that realization has come the hard way. But Steve can recover quickly from his rough start if he brings to bear the engineering mindset he has been developing in his classes. He needs to view his speaking struggle as an engineering problem, and solve it. By focusing some penetrating thought on the matter, he might even gain a level of insight that ordinary public speaking experts lack.

Not wishing to repeat his nightmarish experience, Steve asked for some blunt feedback from his thermodynamics instructor. The response was no surprise: "Your choice of topic was appropriate, but you were clearly unprepared. As you can now see, technical speaking involves far more than being spontaneous and hoping for the best. Good presentations don't just happen — they are planned." But Steve was just beginning to consider all this. He still had many questions. For example, he was aware that presentations could be aimed at a variety of educational levels. Presentations could have drastically different time limitations as well. Some might be intended to persuade, others mainly to inform. What sorts of fundamental considerations underlie all the diverse possibilities? After some pondering of these questions, the general "technical presentation problem" began to strike Steve as a sort of design problem, akin to the complex technical problems he was learning to solve by standard design methods in his engineering courses.

Steve now saw technical speaking as an engaging problem and had a viable way to attack it. We'll return to his epiphany (about how technical presentation can be approached as an engineering design problem) in the next chapter. Right now, let's spend a little time to drive home the importance of being a good technical speaker.

1.3   The Benefits of Being a Good Presenter

Steve is one of the fortunate ones. His "bad experience" occurred in the relatively safe confines of the college classroom. In a professional setting, an engineer who speaks poorly might just (1) come across as unprepared, unprofessional, or even silly; (2) disappoint or annoy important people like mentors, supervisors, colleagues, and customers; (3) mislead others, or at least fail to get hard-won technical ideas across to them; (4) suffer from career stagnation or failure to land an attractive job; and possibly (5) acquire a negative overall reputation as a poor communicator. In short, a poor speaker may have a second-rate engineering career. An ability to speak professionally is a required part of being professional.

Students aren't the only ones who have difficulty speaking in a classroom environment. Consider Steve and Kelly's thermodynamics instructor, Jan. When Jan first became a professor, he had very little experience speaking in front of an audience. The few technical presentations he delivered as a graduate student were certainly no substitute for teaching a real class. Jan thought that being an expert in his field would make teaching undergraduates trivial. Preparing little for his first few classes, Jan found himself bumbling through the material, talking well over the head of the average student. Many students disengaged and stopped coming to class. It wasn't until semester's end that Jan realized he would need to prepare each of his lectures as though it were a technical presentation. He would have to consider the capabilities of his audience, carefully review the material, and silently rehearse the night before. Jan's poor teaching evaluations that semester did not surprise him. However, he was shocked to see comments suggesting he had a poor grasp of his subject! Jan was beginning to learn that preparation and self-confidence (without arrogance) are the keys to winning student respect. He also found that a growing familiarity with the material permitted him to be more spontaneous in the classroom, throwing in anecdotes and teasing out the more subtle and difficult points. Even after two decades of teaching, Jan still spends time carefully preparing each lecture. The payoff is his students' deeper understanding of thermodynamics and their sincere appreciation for his efforts.

Of course, academia is not the only place where the quality of technical presentation is crucial. In industry, the new ideas that ultimately receive support are not necessarily the best ones from a technical standpoint; rather, they are those that are *presented* the best.

Greg and Jennifer have competing ideas for the design of a new educational toy. Greg's concept is much more interesting from a technical standpoint, and he assumes this fact alone will result in his proposal being selected. Meanwhile, Jennifer has done some market research, made a nice mock-up of her design, and carefully rehearsed her speaking points. Much to Greg's chagrin, Jennifer's idea is chosen for funding. Through strategic preparation and skillful delivery, Jennifer made her idea sound more attractive to the audience of decision makers.

Every engineer stands to gain from developing strong technical presentation skills. He or she can expect to (1) please, impress, and influence important persons; (2) come across as trustworthy and confident; (3) provide a valuable service to others, communicating significant technical ideas in a usable way; (4) remain in demand; and eventually (5) build a positive reputation as a superb and effective communicator. Is it worth your time and effort to develop these skills? We think the answer is obvious.

1.4 Getting Started

Try this two-pronged approach. Read the next chapter as soon as possible. At the same time, make an effort to view the presentations of as many talented speakers as possible. Time spent watching and listening to accomplished speakers will repay you many times over. Try to attend the presentations of respected speakers in your organization. There are many other ways to view presentations, both in person and virtually. We give several specific suggestions in the exercises.

You should make an effort to rate presentations as you watch them. Take time to create a *rubric* that evaluates those aspects of the presentations that are important to you. The rubric (aka *scoring guide*) has become a widely adopted category of instruments in education at all levels. Rubrics are used both for assessment/grading and for learning enhancement. The content can be very general, or specific to some aspect of a speaking skill you wish to improve. Here is a very simple example of a rubric for rating a presentation:

	2 points	*1 point*	*0 points*
eye contact	consistent	sporadic	nonexistent
vocal quality (volume, clarity, etc.)	excellent	adequate	inadequate
nonvocal delivery (professional appearance, gestures, confidence, etc.)	exemplary	appropriate	inappropriate
visual aids (quantity, quality, etc.)	highly informative	moderately informative	uninformative
content (scope, relevance, level of detail, etc.)	highly appropriate	appropriate	inappropriate

You should use the same rubrics to assess your own preparation and presentation skills. Do this as you rehearse, and have your colleagues evaluate you as well. You will find the feedback immensely helpful. If you encounter difficulty with some particular aspect of your presentation, make a detailed rubric to help you determine what you need to work on. For instance, if you are having trouble with stage presence, you may want to use the following rubric:

---

eye contact

2 points	*1 point*	*0 points*
Holds attention to the audience. Engages individuals. Does not look at notes.	Looks at audience but also looks at notes. Talks to screen.	Does not look at audience. Stares at screen. Reads notes.

vocal intonation

2 points	*1 point*	*0 points*
Speaks clearly and at proper volume. Varies pitch and pace.	Occasionally stumbles over words. Talks to the screen or away from audience.	Mumbles or is too quiet to hear. Speaks in monotone.

body language

2 points	*1 point*	*0 points*
Poised and confident. Moves rhythmically and uses hand gestures for emphasis.	Seems uncomfortable. Moves in spurts, and gestures at inappropriate times.	Rigid or constantly moving. Gesticulates wildly or always keeps hands at sides.

enthusiasm

2 points	*1 point*	*0 points*
Sincerely cares about the topic. Energizes and engages the audience.	Appears stressed or worried about the talk. Cares, but can't keep audience's attention.	Disinterested. Audience quickly disengages from the talk.

Do not limit your use of rubrics to the act of speaking. You can develop rubrics for the content of a slide, for preparation and rehearsal, and for handling audience questions. We will ask you to write specific rubrics as exercises throughout this book. You can use the checklists provided at the end of each chapter as a guide. We also provide suggestions for developing rubrics below. But be sure to tailor the rubrics to your own needs, and to help you overcome your personal difficulties.

Some Suggestions for Developing Rubrics

Although the references on rubrics listed in our suggestions for further reading (page 140) are all helpful, we highly recommend the book *Introduction to Rubrics: An Assessment Tool to Save Grading Time, Convey Effective Feedback and Promote Student Learning* by Dannelle D. Stevens and Antonia J. Levi. These authors define the term *rubric* as follows:

> At its most basic, a rubric is a scoring tool that lays out the specific expectations for an assignment. Rubrics divide an assignment into its component parts and provide a detailed description of what constitutes acceptable or unacceptable levels of performance for each of those parts.

The constitution of a rubric is thus rather broad and flexible. One classification divides rubrics into *analytical* and *holistic* types. Analytical rubrics are commonly laid out in grid form with *evaluative criteria* listed vertically down the leftmost column and *levels* (sometimes called *benchmarks*) listed across the top row:

	level 1	level 2	level 3	level 4
criterion A				
criterion B		*cell*		
criterion C			*cell*	
criterion D				
criterion E				

The criteria could be the various gradeable aspects of an oral presentation, for instance. There may be more or less than the five (A, B, C, D, E) we have indicated, but for an analytical rubric there should be more than one (as the whole idea of analyzing something is to break it down into components). Three to five criteria are commonly recommended. The levels are typically arranged from best to worst going left to right; they can be worded variously according to your purposes, but among the schemes we have seen are three-level versions such as

high, medium, low

strong, medium, weak

advanced, proficient, basic

excellent, average, weak

proficient, satisfactory, unsatisfactory

proficient, developing, beginning

well done, partially well done, not well done

exemplary, competent, needs work

and four-level versions such as

> excellent, good, fair, poor
>
> exemplary, proficient, developing, emerging
>
> superior, adequate, minimal, inadequate

A purely numerical characterization of the levels is also permissible:

> 4,   3,   2,   1,   0

The grid cells typically contain short descriptions that serve to define the levels. You should strive for clear, concise, and positively-worded descriptions.

See the suggested references (or the web) for information on holistic rubrics as well as various other assessment instruments (checklists, rating scales, and so on).

You are encouraged to search the web for sample rubrics. At the time this book was written, a search for "public speaking evaluation rubric" turned up numerous hits, including a few online rubric generators such as

> http://rubistar.4teachers.org/index.php

1.5   Chapter Recap

1. Careless or incompetent speakers take needless career risks and limit their professional potential.

2. Good technical speech is accurate and appropriate for a particular target audience.

3. Technical presentation has to be far more planned than ordinary conversation or even many forms of public speaking. It certainly involves much more than just being yourself!

4. In industry, the new ideas that ultimately receive support are not necessarily the best ones from a technical standpoint; rather, they are those that are *presented* the best.

5. Trying to bluff your way through a presentation is risky to say the least. You may fool the audience, or you may end up exposed and very embarrassed. Either way, bluffing is unprofessional in a technical setting.

6. The generic engineering design process applies to the design of formal technical presentations.

7. Role modeling after effective presenters is one approach to becoming a good presenter.

8. The rubric is a great tool for evaluating the speaking skill of others and for recognizing and dealing with your own difficulties.

1.6 Exercises

1.1. List some broad purposes for technical speaking. For example, some talks are intended to provide instruction whereas others are mostly intended to motivate.

1.2. Inventory your own attitudes toward public speaking.

1.3. Listen to a sampling of the talks on www.TED.com.

1.4. Attend a public lecture. Take notes on the elements you find particularly effective — and those that you find negative or distracting.

1.5. Listen to or watch the recorded speeches of a favorite politician. What aspects of his or her speaking style do you find particularly effective?

1.6. If you are a working professional, consider joining a nearby Toastmasters International® group. This may give you some opportunities to hone your speaking skills in a friendly environment where your actual job and reputation are not on the line. There is even a virtual club that will allow you to participate via video.

1.7. Most professional organizations give a variety of webinars related to the activities and interests of the organization. Attend a few of these that address topics you find intriguing. Take notes on the speaking styles of the professionals who volunteer their time to deliver these interactive presentations.

1.8. There are many excellent professionally created lecture series that you can purchase or rent to view at home, such as The Great Courses® (www.thegreatcourses.com). Watch some of these lectures and see if you can discern why these speakers are regarded as some of the most talented in their professions.

1.9. If you are a student, pick a favorite professor and try to determine what it is about his or her lecture style that you enjoy. Emulate this in your own talks. If you are already an engineer, pick another engineer or a manager you admire. Ask about their favorite speaking techniques and see whether they have any "tricks of the trade" they are willing to share.

1.10. Engineering programs in North America are accredited by ABET (formerly the Accreditation Board for Engineering and Technology). What does ABET have to say about the importance of oral communication ability for engineering graduates?

1.11. How has globalization affected the need for good, clear public speaking?

1.12. Explore the ethical codes published by professional societies such as the Institute of Electrical and Electronics Engineers (IEEE), the American Society of Mechanical Engineers (ASME), and the American Society of Civil Engineers (ASCE). Do any of the provisions of these codes carry implications about the quality of an engineer's oral communications? Specify.

1.13. Create a rubric to help you evaluate a speaker's ability to hold your interest throughout a talk. Apply your rubric during a formal presentation by someone you respect. Also apply it in a less formal setting, such as a business meeting. Do you need to use different metrics depending on the venue or the type of presentation?

2

Engineering a Presentation

In Chapter 1 we made three claims: (1) technical presentation is challenging, (2) you're going to need to do it, and (3) you should learn to do it well. The rest of the book is concerned with *doing it well*. We also suggested you view technical presentations as projects that can be tackled using the same engineering design principles you have become accustomed to and comfortable with. As we probably needn't remind you, these principles have been wildly successful, producing space shuttles, nuclear power plants, and global positioning systems. We think they can be just as successful in producing technical presentations. So, we'll start by reviewing the basic principles of engineering design, and then see if we can't make our case for applying them to "engineer" a presentation.

2.1 Quick Review of Some Design Concepts

The engineering design process is often depicted as a flow diagram. Figure 2.1 shows its main stages, which are[1]

1. Understand the goal.

2. Conduct research.

3. Generate possible solutions.

4. Choose a solution and implement it.

5. Evaluate, and improve if necessary.

We believe these steps can be adapted to virtually any formal presentation task, and can be your guide when preparing for a substantial speaking engagement.

---

[1] Also see *Engineering Writing by Design: Creating Formal Documents of Lasting Value* by Edward J. Rothwell and Michael J. Cloud, Taylor & Francis/CRC, 2014.

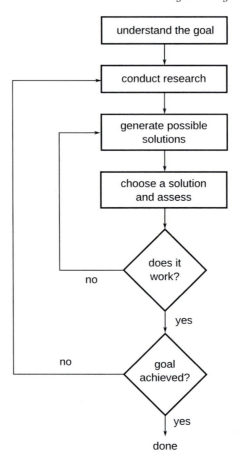

FIGURE 2.1
Flowchart representation of the typical engineering design process.

Mark, a graduate student in mechanical engineering, has finished writing his thesis. It's now time for Mark to defend it in front of his committee and, possibly, an additional audience of invited guests. As the days pass, Mark begins to wonder how he will perform under this much pressure. If he can't summarize and defend his results successfully, the four long years he invested in his research will be at risk.

Suppose some important aspect of your future hinges on a single technical presentation. How can you increase your chances of success? We'll address this question throughout the book. The key is to place this task in an appropriate mental framework, and we choose as that framework the standard engineering design process.

Dr. Stratton, Mark's thesis advisor, reminds Mark of all the hard work that he has put in up to this point. How could he let worries over giving a presentation cancel out so many years of effort? He also reminds Mark that he is beginning a professional life in which formal public speaking is a key component that cannot be avoided. "You're an engineer," says Professor Stratton, "remember that the word *engineer* is derived from the Latin word for *ingenuity*. You'll figure it out, just like you figured everything else out." This switches Mark into an engineering mode: "I have a definite problem to solve here. That problem involves standing up and speaking, but it's still a problem and I can solve it just like I solve other engineering problems." He starts to think about how to apply his knowledge of problem solving to help with his presentation. "How could I *design* a presentation strategy to solve my problem?"

Mark is on the verge of an epiphany. In fact, it's the same epiphany that Steve from Chapter 1 had regarding technical presentations. By using familiar engineering design concepts, he can clear any major technical hurdle, including his thesis defense.

After reviewing the design concepts cited above, Mark responds by drafting a plan as follows.

1. Understand the goal. Mark has 30 minutes to present his principal research results, followed by an undetermined period in which he may have to field some tough questions from his examination committee. His professional future hinges on making a proper impression on this audience.

2. Conduct research. Mark must clearly understand the problem he faces. He must keep two crucial terms in mind: *presentation* and *defense.* Of the many interesting results he obtained over the last four years, which ones are the most significant? Will *all* of the members of his committee understand *why* they are significant? (One member of the committee is a mathematician, not an engineer.) At what level of detail should he present the information? How many diagrams should he show, and how many equations? Should he *read* every symbol in each equation, or merely point to the equation as though it is self-explanatory? Should he display snippets of computer code? Where should he start: with fundamental concepts, or with the crown jewel of his final results? In short, how long is 30 minutes in this sort of situation, and how is it used to one's best advantage? Mark still has contact information for a few former students who recently defended their theses successfully. Should he call them for advice? Are there any books or good websites with information on all this?

3. Generate possible solutions. While Mark can safely settle on just one approach to his presentation, he understands he will have to develop a set of alternative approaches to the question-and-answer phase. Depending on their moods that day (and, of course, on how well Mark performs in the presentation phase), the audience may turn out to be friendly or somewhat hostile. Furthermore, it is possible that some razor-sharp outsiders may attend and ask unforeseen questions. Rather than planning to generate workable responses to just one scenario, Mark follows the standard design process and plans to generate *multiple* potential responses and modes of response.

4. Choose a solution and implement it. Evaluate, and improve if necessary. For Mark, this choice will have to be made partially in real time. This is one aspect in which speaking before a live audience differs from writing. He can settle into a well-planned presentation, but he must be ready for a number of possibilities during the question-and-answer phase.

Mark will have to prepare slides, rehearse, solicit feedback from others, evaluate their comments, rehearse again, try to obtain feedback again, and so on. Some feedback will come during the talk itself; in this case Mark will have to re-evaluate and refine "on the fly."

By adapting Figure 2.1 to his purposes, Mark has generated a basic but essentially complete roadmap for himself. He knows that by following it he can successfully defend his thesis, and for the first time feels confident about speaking in public. Mark thinks of all the hard work he has put in during graduate school. He senses that he might finally hear those magic words, "Congratulations, Mark. You passed!"

As an engineer, you can adapt your speaking tasks along these same lines. Whether you're preparing a briefing for your manager, a senior project presentation, or even a dissertation defense, a design-based approach will help you prepare, control nervousness, and actually enjoy giving your talk. So let's delve more deeply into some of the fundamental aspects of the design process.

Top-Down and Bottom-Up Approaches to Design

Given a design task, we may decompose it into subtasks and then break down those subtasks, continuing until we are satisfied with the level of decomposition. This idea is the basis of the *top-down* design approach (Figure 2.2).

For example, an engineer might dissect an air conditioning system into a compressor, a blower, an accumulator, an evaporator, and a condenser. The compressor in turn could be decomposed into a crankshaft, connecting rod, piston, discharge valve, and discharge line. Lower level subsystems are then designed and assembled into subsystems at the next higher level until the whole system is complete. This approach — analysis followed by synthesis —

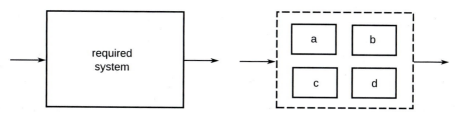

FIGURE 2.2
Notion of top-down design. *Left:* Top-level view of system to be designed. *Right:* Breakdown of required system into subsystems, shown as blocks **a**, **b**, **c**, **d** here. In the analysis phase, each subsystem is further decomposed until the concrete design stages are reached. During the synthesis phase, the completed subsystems are assembled into the required system. The basic idea is divide and conquer.

is strongly advocated as an engineering design approach and can be easily applied to the design of significant formal presentations.

Mark decides to take a top-down approach with the presentation phase of his defense. He begins by realizing that he will need a short introductory statement, the main body of the presentation, and at least a brief concluding statement before the questions begin. He ponders the various possible ways to introduce himself and his research area, looking for something efficient but not terse or off-putting to a mixed technical audience. Estimating that 25 minutes will remain for the meat of his presentation, he turns to the principal result of his research, a new relation pertaining to turbulent fluid flow near an irregular boundary. Estimating that it will take 5 minutes to state the result and its immediate corollaries, Mark realizes he can allocate 20 minutes to background material that supports the main theorem. Continuing to decompose this available time into five-minute segments, Mark sees that he can cover just four of the most important background topics.

In contrast, consider the less focused approach of the "idea engineer" who is hired to come up with new concepts, gadgets, or procedures. He might fool around with an electronic circuit with no particular end goal in mind, and find that he can make it do something unexpected. He couples the new circuit into a system he has designed previously and improves the efficiency of the system dramatically. He can now put the system to use in situations where cost would have been prohibitive, and through this sequence of unguided events completely revolutionize the industry. The design process described here does not fit the top-down category but is a design approach nonetheless. While *bottom-up* design is less conventional than the top-down paradigm discussed above, it allows more room for creative adaptation and often leads to major breakthroughs (and an equal number of complete flops). Many people, including speakers and writers, prefer to work in bottom-up mode. When you find yourself stuck early in the planning of a major speaking task, try shifting into low gear and dumping your material into an electronic document so you can play with it. Move things around, reword them, insert dummy placeholders, etc., until organization begins to emerge. Small successes can lead to larger successes, and don't be surprised if a sensible presentation evolves.

> Mark realizes he cannot "plan" the question-and-answer portion of his defense. What he *can* do is try to anticipate the most difficult questions that are likely to arise. Switching to bottom-up mode, Mark begins to collect answers to potential questions as he thinks of them. Several days later, he has accumulated a file full of efficient (and sometimes even evasive) answers. He composes several extra slides that anticipate the questions, and appends these to the end of his presentation in case they are needed.

As we see, a full design process can take the form of a *hybrid* approach with both top-down and bottom-up elements. Engineering is about what works.

A crucial element of both approaches is *iterative improvement*. Iterative means repetitive; it refers to an approach where successive attempts are made, each one building on the previous one. You start with your chosen solution and take successive cuts at improving it until it meets the standards for completion (recall Figure 2.1). This is how you should approach the preparation of your technical presentation. You will also use iterative improvement to polish your delivery through *rehearsal*.

Notion of Concurrent Design

Imagine an engineering team charged with developing a new entry in a successful product line. They work very hard for several months on a specific design, only to learn that it cannot meet industry standards for post-consumer disposal. This is a risk of the *sequential design* approach. First the engineers work in isolation and only later get input and advice from other professionals who have an impact on the success of the project. Contrast this with a design team that integrates experts in marketing, manufacturing, disposal, and so on, at the outset of the project. This approach is called *concurrent design*.

You probably won't have the luxury of assembling a team of experts to help you prepare for a speaking engagement. Nonetheless, the basic idea of getting information and early feedback certainly applies. If you must speak in an auditorium, for example, it may be wise to practice in a similar room before the big day arrives. Rooms differ widely in their lighting and acoustical properties. A huge room may even require the use of a microphone and sound system. Early awareness of such factors can be quite helpful. You may also seek feedback, rehearsing in front of people in (or close to) the target audience and asking for reactions. Their suggestions, if addressed early in the preparation process, could prevent headaches later on.

Having planned the presentation phase of his defense, Mark rehearses and gets feedback from the other students in his group. Fortunately, one of the students who developed part of Mark's experimental procedure manages to catch an error in Mark's explanation. Mark corrects the error and thinks, "Wow, I'm glad I didn't make that blunder in front of Dr. Stratton!" The next day, Mark rehearses in front of the professor and receives plaudits for an excellent practice presentation. Mark is deemed ready to schedule his defense.

2.2   Framing the Goal of Your Technical Presentation

When engineers speak professionally it is primarily to communicate thoughts and ideas rather than, say, to emote or to entertain. This is not to say that technical speaking must be robotic or sterile; it's important to have one's passion for a topic shine through in a presentation. But the primary goal of technical speaking is to inform. We must tell our listeners *who, what, where, when, why,* and *how*. In the process, however, we cannot afford to forget the listeners themselves. *We must keep the backgrounds, needs, and purposes of the audience firmly in mind at all times.* We are speaking to inform, not simply to "talk about" equations and diagrams. On the other hand, we are not aiming for "perfection" (whatever that is). We are simply trying to do a job and do it well. An engineering design process does not produce a perfect product, and a process to prepare you for public speaking will not produce a perfect fifteen-minute talk. Intelligent compromise is basic to all practical engineering activity.

Let's reiterate. As an engineering professional or student, you cannot avoid technical speaking. Eventually you will need to speak in front of a class, pitch an idea to management, or speak at a conference. Fortunately, you can make your preparation less daunting by applying the engineering design process. Recall that the first step in that process (Figure 2.1) is to thoroughly understand your goal. As a public speaker, your fundamental goal is effective communication. More formally,

Your goal as a technical presenter is to communicate information to an appropriate target audience.

In other words, some body of information currently resides in your mind and must be made available to the minds of the target listeners. There are two main issues here.

1. How does the information reside in your mind? How did it get there, and what forms does it take? Which portions of the body of information exist in your mind as visual impressions ("pictures"), or as abstract concepts, equations, quotations from authorities, etc?

2. Who is the target audience? What are the attributes common to those people at whom you will aim the presentation?

If your knowledge of a topic isn't clear to you, or if it isn't clear who you're trying to communicate with, your chances of producing an effective presentation will be small. Not convinced? Picture an engineer who is to design a circuit to control the movement of a mechanical device, but doesn't know what the device is. How will he proceed with the design if he doesn't know what range of motion the device will experience? Is the movement longitudinal or rotational? What weight must be moved? How fast must the movement be? Without this information, the engineer can't possibly begin to design the system. And without knowledge of your topic and the audience, you can't begin to design a presentation.

You have probably seen a presentation where the speaker seems to have "thrown together" a bunch of slides full of words, figures, equations, headings, etc., and flashed them past the audience. He may not have even considered the makeup of his listeners, and as a result you probably felt under-served. Perhaps the speaker had only a short time to prepare. Perhaps he thought:

I've just got to make it through this presentation and keep my job.

This is an untrained approach to speaking. You should strive for better. Your audience deserves better.

In successful technical speaking, as in successful engineering design, *form must match intended function*. That all-important match will not likely happen automatically or at random. Remember the goal statement:

Your goal as a technical presenter is to communicate information to an appropriate target audience.

In order to communicate something of technical value to others, you must be clear about what you know and how you know it, understand the characteristics of the intended audience, and consider how to best transfer your knowledge to that specific group of listeners. Let's examine the first two of these aspects in greater detail.

2.3   How the Information Resides in Your Mind

Engineering information may reside in your mind in a variety of ways. It may take the form of a mental image, like a mechanical drawing, or it may exist as an abstract concept such as "pressure" or "density." It may be formed from other sense experiences, such as the sound of a failing bearing, or the smell of a chemical, or it may be the purely symbolic mathematical expression $f = ma$. It might even be best described as a sequence of words, such as in rules of thumb: "smaller loops radiate less," or "less contact area means less friction." (Be careful with such phrases, though, since they lead to lazy thinking and are often misleading or even wrong.)

It's not our purpose here to adventure into psychology, and we're not worried about how information might be stored in the brain. But it's important for you to understand how you, yourself, view the essential aspects of a topic if you hope to make a meaningful connection with an audience. Once you understand your relationship with the information you want to present, you can think about how to pass the information across to the listeners. Work hard to maintain an awareness of this relationship throughout the process of planning any presentation.

> **Example.** Melody's senior project involves filtering noisy radar signals to improve detection rates. She tries to think of a good way to describe the effects of noise to other students, and realizes that she most strongly associates noise with an audio signal. She thinks that perhaps it is easy for students to "hear" noise, even if they don't have a good mental picture of the generic concept of noise. So, she applies her filters to an audio signal and plays the sound files to her audience. "Think of the radar signal as a sound," she says. "When there is noise, it's like an added hiss. When we lowpass filter the sound signal, we get rid of a lot of the hiss." The audience immediately understands the impact of the filter on the signal, and the demonstration creates a nonverbal analogy in their minds that they can apply to the general concept of noise, just as Melody does.

You're the expert on the subject of your presentation. If you have found a helpful way to understand some aspect of a topic, then why not pass this along to the audience? If you think a topic is best understood as a sequence of drawings, then use drawings. If it's best to use equations, use equations (carefully, of course).

> **Example.** A table can be a clear way to present things. Consider this table, which exhibits properties of some types of electrical capacitors:

type	voltage rating	series resistance	RF frequency range	market share
ceramic	high	low	high	46%
mica	high	low	high	< 3%
tantalum electrolytic	low	high	low	12%
aluminum electrolytic	low	high	low	22%

Would this data be as clear if it were presented as running text on your slides? Compare the table to:

> Ceramic capacitors have high voltage ratings, low series resistances, high RF frequency ranges, and a 46% market share. Mica capacitors have high voltage ratings, low series resistances, high RF frequency ranges, and market share of less than 3%. Tantalum electrolytic capacitors have low voltage ratings, high series resistances, low RF frequency ranges, and a 12% market share. Aluminum electrolytic capacitors have low voltage ratings, high series resistances, low RF frequency ranges, and a 22% market share.

Clearly, the data is much easier to access using a table.

2.4 Your Audience

Effective speakers put their audiences first. As an engineer, you are speaking to inform — not to bluff, dazzle, impress, or enchant, and certainly not to confuse or frustrate. Think about your potential listener: try to envision him or her.

1. What is the listener's background? Is he an accomplished expert with education appropriate to the topic? Is he a nontechnical but powerful decision maker who expects you to display fluency and confidence?

2. What are the listener's purposes? Will she simply want the bottom line regarding the topic? Is she after a much deeper view? Is she evaluating you, or in some other way making a decision affecting your future?

3. What is the listener's likely level of understanding? Is he an undergraduate student? Is he a member of your doctoral committee? Is he a graduate student outside your technical area?

We can't overemphasize this fact: the audience is central to all considerations regarding a presentation. Planning a talk without understanding the target audience is like designing a product without understanding the customer.

> **Example.** It's your job as a speaker to know not only what the audience wants to hear, but what it needs to hear. Zoë spent a long time preparing a talk to her engineering staff on how to implement new design protocols for the high-power systems manufactured by her company. After Zoë outlined the changes, several staff members grumbled, and one brave employee spoke up. "Wow, this is going to make our lives a lot harder. Why the heck do we have to to do this? It seems like this company is always introducing meaningless changes just to make us suffer." Zoë realized she should have led off her talk with some background on why the changes were needed. Her engineers had spent a great deal of time tweaking the old protocols and had become efficient at implementing the designs; she should have known how sensitive they would be about the changes. "I know this is going to be a lot of trouble," she said, "but it came from a recent overhaul of government regulations and we really have no choice in the matter." It took a while to calm her staff down, but eventually they were convinced that they had no choice in the matter.

There are many other useful analogies between technical presentation and engineering design. Consider:

technical presentation	engineering design project
audience	customer
effective speaking conventions	engineering standards
preparation time and effort	project cost
brevity and conciseness	product efficiency
clarity	product effectiveness
rehearsal	test run
critique	customer feedback

Like any consumer product, your presentation must address your customer's needs. It must adhere to expected conventions such as those of English grammar. It must economize the listener's time, energy, and patience. Finally, it must fall within budget in terms of your own time and energy as a busy engineer.

2.5   Other Aspects of Situational Awareness

Technical presentation is complicated. There are other things to consider besides your topic and the listener. Here are additional questions to ask during the planning stages.

1. What is the time limit? Technical presentations run the gamut from ten-minute talks to multi-day seminars. Time length is obviously a crucial parameter to identify at the outset.

> **Example.** Sandra was worried that she wouldn't be able to fill an entire 20-minute time slot at the regional chemical engineering conference. It took her by surprise when, 22 minutes into her presentation, the session chair nervously pointed at the clock. This threw Sandra off and created a discontinuity in her presentation when she realized she had to skip to a severely abbreviated conclusion statement.

2. How will the audience be positioned? How and where will they be seated (in particular, how far away from the speaker)?

> **Example.** George expected his nifty working demo unit to carry the day at senior project presentations. He didn't anticipate having to deliver the presentation in a large classroom. George held up his palm-sized box while people in the back row squinted and clearly had no chance of reading the tiny display. After the presentation, George realized he should have created a short video to show on the projection screen instead. Those who were interested in holding the actual demo unit in their hands could have come up after the presentation to do so.

3. What presentation equipment is available? You don't want to prepare transparencies only to find no overhead projector available. If electronic projection is to be used, be sure you have used the correct software to prepare the presentation. Also check that the version of the software is compatible.

> **Example.** In the middle of his conference presentation, Eric was horrified when the equations he so carefully prepared using the required presentation software appeared as a sequence of strange characters resembling mailboxes, flowers, and computer monitors. It turns out that while he was careful to use the prescribed software, he didn't use the correct version. He probably took away more from the presentation than the audience by discovering how it feels to wing a technical talk without crucial visual aids.

4. What are the other aspects of the speaking environment? What is the acoustical situation? The lighting situation? Will you be competing with external or internal noise sources? Where will you appear in the speaking program?

Example. Lisa had presented in large rooms before. This one was different, though, and Lisa didn't understand that until she started speaking. The room was narrow but very deep and, most importantly, *carpeted* with rows of fabric-covered chairs, thick wallpaper, and a porous drop ceiling. Her words were largely absorbed before reaching past the first few rows, and people sitting in the back struggled to hear. Lisa found herself distracted from her topic, trying to calibrate her vocal volume to the situation. Halfway through her talk, an audience member in the front row boldly asked Lisa why she wasn't using the microphone that sat over on the podium. Terribly embarrassed, Lisa muddled through the rest of her talk and sat glumly in the back of the room after finishing.

2.6 Checklist: Engineering Your Presentation

- ☐ **I understand the goal of my presentation**
- ☐   I understand how the relevant information resides in my mind
- ☐   I fully understand the target audience
- ☐   *I know the background of the audience*
- ☐   *I know the purpose of the listeners*
- ☐   *I know the level of understanding of the listeners*
- ☐ **I have completed the research necessary to generate solutions to my speaking task**
- ☐   I know the time limit
- ☐   I know the limitations of the venue
- ☐   *I know how the audience will be positioned*
- ☐   *I know what presentation equipment will be used*
- ☐   *I'm aware of the acoustics and lighting*
- ☐ **I have generated several potential solutions to my speaking task**
- ☐ **I have evaluated the potential solutions and decided on the one that best meets my needs**

2.7   Chapter Recap

1. The generic engineering design process applies to the design of formal technical presentations.

2. You can plan a presentation via the customary divide-and-conquer (top-down) approach with iterative improvement.

3. As engineers, we know that it's hard to solve a problem without understanding it first.

4. Your goal as a technical speaker is to communicate information to an appropriate target audience. In other words, a presentation is intended to communicate what you know to an interested, reasonably prepared listener.

5. Producing a formal presentation *just* to meet a deadline or attain some other reward is subject to the old rule of *garbage in, garbage out*.

6. If given the latitude, pick a topic you're really interested in. Your enthusiasm will contribute much to the quality of your presentation.

7. It helps to think about what you know, and how you know it, before trying to present it to someone else. Doing so might even help you become a better subject-matter expert.

8. Good technical speech is accurate and appropriate for a particular target audience. It's essential to consider the target listener's background, purposes, and maturity level. This is especially the case with mathematical maturity.

9. If you dislike formal speaking but enjoy pleasing your customers as an engineer, then think of each audience member as a customer.

10. Gather pertinent information on the target venue well before the speaking event. Study environmental factors such as acoustics, lighting, temperature, humidity, seating, and outside noise. Test the projection equipment if possible, making sure it works with your own laptop computer.

11. Get feedback early and often. (And, as a professional courtesy, provide feedback to others who seek help preparing for their speaking engagements.)

12. Thinking like an engineer is not just a paradigm for giving a talk; it is also a framework for evaluating the formal talks of others. Engineers must review and critique the information in many oral presentations.

2.8 Exercises

2.1. What types of design constraints do engineers routinely face? List as many as you can.

2.2. Pose a simple problem and generate at least three alternative solutions.

2.3. If an idea dawned on you in a flash of insight, must you still lay it out systematically for the listener? Why or why not?

2.4. Todd just embarrassed himself in his control systems course. His total improvisation approach to the semester oral presentation fell flat and resulted in a low grade. He is disoriented and demoralized. However, Todd still has one week to prepare for a similar presentation in his electromagnetics course. What advice would you give him?

2.5. Choose a topic within your area of knowledge and consider how you'd explain it to a layperson. How would you break it into manageable chunks? Would you have to further decompose some of these chunks to make them understandable to the listener?

2.6. Make a list of signpost words that could serve as headings or subheadings on your own presentation slides. Some possibilities are as follows:

- Approach
- Demonstration
- Explanation
- Limitations
- Purpose
- Review
- Validation

2.7. Organize some technical information in table form. Choose any topic of interest.

2.8. Explore your building or campus and look at the various rooms in which presentations occur. Stand at the podium and see how well your voice projects. Take notes about the available presentation equipment and about how to make yourself heard in each room. Do you find significant differences between rooms?

2.9. Construct a rubric to evaluate how well you have engineered your presentation. You may wish to use the checklist from Section 2.6 as a guide.

3

Designing Your Presentation

Here we consider the basic structure of your presentation. We offer some simple ideas and guidelines to get you started regardless of the purpose or content of your talk.

3.1 Does Your Presentation Need a Structure?

Of course it does, and it's important that the audience senses that it *has* one. Your listeners are entrusting you with their time, energy, and attention — they need to know that care and forethought lie behind their listening experience. A presentation without discernible structure can be a disaster.

> David, an undergraduate senior, opened his capstone design presentation with a detailed electrical schematic diagram:
>
> > I spent most of my time in this project interfacing the microcontroller chip to the various sensors and display units. Here, in the upper right-hand corner, is an example of the many subcircuits I had to design. The value of current-limiting resistor R_{45} was chosen as 4.7 kΩ because ...
>
> The diagram, with all its complexity and organization, was certainly impressive. But no one really knew what David was talking about — even the electrical engineers in the audience were perplexed. What was the point of his system? Was it part of a clothes dryer? A robot? A game? What environmental variables were the sensors there to monitor? As these questions floated through their minds, the listeners got increasingly lost and, eventually, annoyed. They began to feel that David was talking *at* them rather than *to* them. Soon the audience just gave up — David had lost them completely.

David assumed that the most challenging part of his project would carry the most interest for his audience. This may have been true, but he never gave his listeners a chance to find out. His omission of a suitable introduction produced an immediate sense of perplexity, and his presentation fell flat. A good presentation *always* begins with an effective introduction.

Betty was assigned to give a half-day seminar on acoustics. Taking the stage, she dropped a heavy chunk of scrap metal onto the floor — *clang*! Then everything went silent.

That was quite a noise! And do you know what noise is? It's *sound*. So take your fingers out of your ears and let's talk today about sound.

Betty's dramatic introduction had grabbed her listeners; from that moment onward they hung on every word. Several years later, people were still remarking on how Betty's talk began that day. By quickly engaging and then holding her audience, her presentation became much more than a successful company seminar — it became a memorable learning experience.

The cases of Betty and David show that the introduction is crucial to the success of a technical talk. Unlike David, Betty gave the listeners a bit of context and something familiar they could build on. Her introduction served to ease the audience into her subject. The basic role of a conclusion is similar: it provides a way to ease the audience back out.

Polly was thrilled to navigate the main body of her presentation without a glitch. She clearly had her listeners interested and informed about broadband antenna design. But then Polly simply stopped talking. After 20 seconds or so, the audience starting wondering whether she was OK. Was this a "dramatic pause" for emphasis? Emphasis of what? One listener turned to look toward the back of the room: had someone entered and silently signaled Polly about something? In fact Polly had finished her talk without a concluding statement and was now waiting for a reaction from the audience. But she had simply left them hanging. Was she done? Should they applaud? Was she inviting questions? Should they just exit the room? (Wouldn't that be rude?) As people shifted uncomfortably in their seats, Polly said awkwardly "I'm done now."

Less egregious but more common is the speaker who just steamrolls through his material until suddenly reaching a "Questions?" slide. Although such a slide signals an audience that a presentation has ended, the transition is too abrupt and a proper sense of closure is lacking.

In broad terms, every technical presentation needs three main phases: an *introduction*, a *main body*, and a *conclusion*. Convention alone leads an audience to expect each of these components; when any is lacking, confusion results. Let's pause and consider why structuring a presentation around these three phases is so effective.

Smooth Beginnings and Endings: The Two-Funnel Picture

Those who have heard accomplished technical speakers in action may have noticed a pattern. A typical presentation will contain some real scientific meat, but the speaker won't start or end the talk with that material. Instead, she will have to contend with an audience approaching her talk from many different directions and initial mindsets. Some listeners may be seeking specific information, while others are attending out of plain curiosity or coercion by their bosses. A proficient speaker wants to give all these listeners a chance. So, does the following sound like a good way to start a technical talk?

> Let's take a look at the microcontroller chip, shown here connected to several sensors and its output buffer.

Ugh. Sure, somebody in the audience might adore microcontrollers, but chances are that most listeners would want to hear about the overall system first. What does it do? Why is it important to our company? Is this for a new product, or are you trying to improve an existing system? Without some level of context, the average listener is going to look for any excuse to tune you out or sneak out of the room. Let's try again:

> Good afternoon. My guess is that nearly everyone here has experienced the pain and inconvenience of having to coordinate sensor input to the engine control unit ...

Wow. Wouldn't you want to hear more about any proposal for decreasing your pain and inconvenience? We sure would.

A broad, attention-getting approach like this will be much more appealing to a greater share of the audience. Does it take sensationalism or hyperbole to "grab" a casual listener? No, not if the listener has an interest in what you have to say. But you do have the difficult job of finding a way to deal with a potentially diverse audience.

Imagine the start and end transitions of your presentation as a pair of funnels (Figure 3.1). The introduction is used to funnel a variety of listeners into your talk. The meat is at the center of the presentation. The conclusion funnels the audience back out, leaving the listeners comfortable and ready to engage with the speaker.

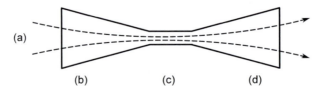

FIGURE 3.1
Two-funnel model for friendly speaking: (a) listeners arriving from various angles; (b) funnel them in; (c) deliver the information; (d) release them gently back into the wild.

The two-funnel picture is an example of conventional structure. It is effective because it engages the audience, accomplishes the goal of presenting the factual information, and then disengages the audience in a way that leaves them satisfied. It is also a structure most audiences are familiar with and have come to expect. So you really can't go wrong if you sandwich your technical meat between an introduction and a conclusion, *as long as you design each part properly.*

3.2   Designing the Introduction

The design of your introduction depends on what you are trying to accomplish. A "well-designed" introduction may have several goals, but they all come down to preparing an audience to receive your information. In their book[1] *Public Speaking: An Audience-Centered Approach*, Beebe and Beebe mention multiple things that could be accomplished during the first moments of your talk. They include

1. getting the audience's attention,

2. introducing your subject,

3. giving the audience a reason to listen,

4. establishing your credibility as a presenter, and

5. previewing your main ideas.

The structure of your introduction, or your approach in delivering it, may also take many forms. In their book *Speaking: Back to Fundamentals*, Logue et al. provide a very helpful list of general forms your introduction could take. The introduction could be written with

---
[1]See "Further Reading" for the full references for all the public speaking books mentioned in this chapter.

1. reference to the occasion;

2. reference to a recent incident;

3. reference to the preceding speaker's remarks;

4. reference to the subject;

5. the use of humor;

6. a statement of your specific purpose or thesis;

7. a preview of points.

Let's give some examples showing how items from these two insightful lists could be selected and combined in different ways. We'll stay with David's capstone design presentation; after all, he'll have to repeat it next semester.

Example. (refer to the occasion; establish credibility)

I am pleased to talk about my capstone project here today. This presentation culminates seven months of work, which began with the realization that ...

Example. (refer to recent incident; get audience's attention)

On June 14, just a few weeks ago, the eyes of the entire world turned momentarily toward a horrible accident in which a hot air balloon touched a high-voltage power line ...

Example. (refer to preceding speaker's remarks)

It is certainly fitting that I come right after Bill Smith today. I was especially intrigued by his observation that many recreational activities lack sufficient provisions for safety ...

Example. (refer to the subject)

Safety systems are essential aspects of modern life. We are surrounded by sensors that monitor our environment ...

Example. (state specific purpose or thesis)

> Today, I am here to give you a concise summary of my capstone
> design project ...

Example. (preview points)

> Good afternoon. I intend to cover the following aspects of my
> project today ...

You'll notice we omitted an example using humor. Although effective in certain instances, humor is also risky. Many a speaker has unintentionally alienated a listener through an insensitive remark. Humor should be reserved for the accomplished speaker who knows his audience well and has an exceptional grasp of the speaking situation. If that's not you, you're safer going with one of the other approaches listed above.

3.3   Designing the Conclusion

Let's skip over the main body of your presentation for now and consider the conclusion phase of your talk. As with the introduction, the way you design the conclusion depends on what you seek to accomplish with it. The purposes of a conclusion can include, according to Beebe and Beebe,

1. summarizing what you said,

2. emphasizing the main idea one last time,

3. motivating the audience to do something after the presentation, and

4. providing the listeners with a sense of closure.

You remember Polly, our speaker whose plan was to conclude her presentation by simply falling silent. Let's consider some examples of what she could have done differently.

Example. (summarize)

> I will conclude by summarizing my main points. First, broadband
> antenna design requires a fundamentally different approach from
> narrowband antenna design ...

Example. (emphasize main idea)

Let me conclude by emphasizing that the key to broadband antenna design is Rumsey's principle: an antenna is frequency-independent if its shape is specified purely in terms of angles ...

Example. (motivate the audience)

To conclude, I hope I have shown you why broadband antennas can give you a competitive advantage when multiple system requirements are ...

Example. (provide closure)

What would I like everyone to take away from this talk? The main thing is that broadband antennas constitute a unique area with many interesting questions still open for investigation. It has been my pleasure to speak briefly about this area today. Thank you so much for your time and attention.

Notice how Polly could use keywords to signal the audience about her intent (e.g., *Let me conclude by*). Next time Polly will leave no doubt that she is bringing her presentation to a close. Her decisive finish will seal the hard-earned impression that she is competent and in control of the situation.

There are many ways to indicate to the audience that you are at the end of your presentation. Some speakers display a

Thank you!

or

Any Questions?

slide after their concluding statement. This confirms that the presentation has indeed ended (which is a good idea when multiple slides were used during the conclusion phase). The speaker could also accomplish this verbally with

I welcome any questions you may have.

We will talk about the art of fielding audience questions in Chapter 6.

3.4   Designing the Main Body of the Presentation

Preparing the meat of your presentation obviously requires the most thought and the most work. As with your introduction and conclusion, the design of your main body depends on what you are trying to accomplish. Engineers give presentations for many reasons, so it's tough to give an exhaustive classification. Certainly you've experienced many of the following, either as a speaker or a listener:

briefing	status report
proposal	progress report
conference paper	final report
training session	thesis defense
product demonstration	job interview talk

Some sort of hybrid presentation is also possible. We can't tell you what to say in all these situations; there are no hard and fast rules. That's why our thesis

> A technical talk is something you should *design* to fulfill a set of requirements.

is so effective. And as we stated in the last chapter, you must understand those requirements early in the design process.

Example. Chris was assigned to give a briefing. This is a specialized type of presentation on which much has been written and many good suggestions are available. By skimming a few books and websites, Chris learned that a number of briefing formats were in vogue. A little thought led him to pick one of these formats for his purposes.

Still, it is often useful to have access to broad principles. In their book *Public Speaking*, Osborn and Osborn give a useful list of general speech design patterns. The material in a presentation might be ordered according to

1. how items are arrayed in space (*spatial design*);

2. how events occur in time (*sequential design*);

3. how your subject naturally or traditionally divides itself (*categorical design*);

4. how your subject stands in comparison or in contrast to something that the audience already understands (*comparative design*);

5. how certain conditions produce other conditions (*causation design*).

These are tried-and-true organizing principles, and one of them may suggest itself as a good choice for your purposes. Or perhaps you may wish to employ a hybrid approach.

> **Example.** Kevin was assigned a 10-minute presentation slot to describe a new geartrain. The spatial arrangement of the gears was clearly the best way to organize his talk. Kevin decided that after a one-sentence introduction he would simply work through his diagram from left to right.

Kevin made exclusive use of *spatial design.*

> **Example.** Scott was given 35 minutes to explain a new shut-down procedure. The situation necessitated that things be handled in a strict order: if someone cut connection B15 before moving switch A to the off position, the entire system could be damaged. Scott basically had to order his information chronologically.

Scott made exclusive use of *sequential design.*

> **Example.** Brian's semester in-class presentation would require him to talk about the principles of antennas. Because the ideas involved were relatively abstract, Brian settled on a three-pronged attack. He would start with a comparative approach, exploiting an analogy between electromagnetic waves and water waves. He would then switch to a causative approach, describing how electromagnetic waves are produced by antennas. Finally, he would settle into a categorical approach, presenting three commonly accepted classes of antenna structures.

Brian's situation was more complicated. He made effective use of a hybrid approach.

Persuasion: Always Part of the Picture

This book deals primarily with factual presentation. Even so, it is important to motivate the audience so they have a reason to pay attention. Why did you do the work you are describing? Why should they care? These questions are often answered immediately following the introduction. We can easily envision (if we haven't already experienced) the grumpy engineering manager — let's call him Fred — who rejects a sound idea or proposal simply because it is "not sufficiently well-motivated." Weeks spent preparing a technically impressive talk are of little use if Fred won't give it a serious listen. Worse yet, Fred may get up and leave halfway through the talk, before we even reach the crown jewel of our technical thoughts.[2] In other cases, the purpose of a talk may be specifically to get attention or obtain funding. Assuming our idea *is* technically sound, how can we give it a better chance in the constant clatter of a very competitive marketplace?

The effective use of English falls under the heading of *rhetoric* and many books on persuasive speaking are available. Rhetoric has been an important part of discourse since the days of the ancient Greeks; for many centuries it composed one third of the "Trivium" of classical education (the other parts being grammar and logic). We do not attempt a detailed description, but merely offer a few suggestions that you may have occasion to work into a talk.

1. You must quickly gain the listeners' interest. Otherwise they may stop listening and direct their time and energy elsewhere. Assume you have an extremely narrow window of opportunity. Think of the classic *elevator pitch* in which the speaker has only one or two minutes to sway the listener. This brings us to ...

2. You must somehow address a need for the product. The product could be an idea or your plans for a new engineering system. Perhaps you're speaking to inform people who need to know about a subject. Maybe you're speaking to advance an idea that will solve a significant technical problem. The point is that you must communicate these things, possibly to a skeptical and distracted listener. Don't expect a complex technical talk to act as evidence of a need for its own existence.

3. You must offer a solution. You must argue (or at least state) that the product addresses the established need.

4. You may have to differentiate your product from its competitors. That is, you may have to argue that the product addresses the need better than other available products.

For the listener, these things may constitute the only available reasons why your product should be seen as useful and worthy of consideration.

---

[2]Don't be too hard on Fred. He's a busy guy, and has to suffer through mediocre presentations all the time. Moreover, Fred's ideas and proposals are often treated the same way by his own resource-conscious superiors.

Example. Before Vinny could make serious plans to pursue the automotive design of his dreams, he had to pitch the idea successfully to his superiors. Vinny was an engineer, not a salesman. On the other hand, no salesman could possibly grasp the technical ideas on which Vinny's dream design was based. So he had to step out of his comfort zone and sell the idea himself. When the time came, Vinny was ready. With a tightly planned presentation he got the attention of everyone in the room, convinced them the market was ripe for his idea, dazzled them with brilliant artwork, and then closed the deal by establishing the technical and cost feasibilities of the design.

3.5 Getting a Plan on Paper

One way to start is with an outline. A typical outline looks like this:

1. Topic
   (a) Subtopic
       i. Sub-subtopic
       ii. Sub-subtopic
       iii. Sub-subtopic
   (b) Subtopic
       i. Sub-subtopic
       ii. Sub-subtopic
2. Topic
   (a) Subtopic
   (b) Subtopic
3. Topic
   ⋮

While most people are familiar with formal outlines, they aren't very helpful when you are unsure of where to begin. Perhaps you have some general ideas, but they are jumbled or disorganized. Maybe you have too many undifferentiated ideas, and you need a method to tease out some structure. In these cases, you may find that an alternative method works well.

The Mind Map for Technical Presenters

If an outline doesn't seem to be working for you, try the following. Grab a piece of paper (or your tablet computer) and write the topic of your presentation near the center of the page. Draw a dark or double oval around the topic. Next, think of some short phrases that describe the main aspects of the topic. Write these near the main topic and draw ovals around them. Now connect these ovals with the main topic using solid lines. Repeat this process treating each of the secondary topics the way you just treated the main topic. Think of phrases that describe the main aspects of one of the secondary topics, write those down, draw ovals around them, and connect them to the secondary topic. Do this for all of your secondary topics, and then progress another level, if needed, until you have exhausted all of your ideas. You should have a very nice visual picture of how all of the ideas flow from one another, and you can add interconnections between any of the ovals to help you organize how various ideas fit together. See Figure 3.2.

During the process you may sense a need to indicate other connections between existing ovals. Do this with dashed lines. This can help you continue expanding your mind map even with subjects that are a bit tangled by nature. One such dashed line is shown in Figure 3.2.

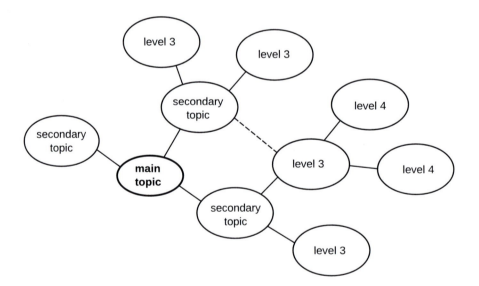

FIGURE 3.2
Notion of a mind map. The picture begins with the *main topic* oval, after which various offshoots suggest themselves.

Example. Landon was preparing to pitch an idea for a new product to his design staff. The concept was still a bit fuzzy in his mind, so he was uncertain how to lay out his ideas. He hoped that a mind map might help him organize his thoughts, and started putting things down on paper. After a bit of work he noticed that the map was quite lopsided, with several single bubbles, and a few bubbles with extensive branching. This phenomenon is fairly common — at the nascent stage, mind maps are often sketchy and dominated by a few topics. As he continued to think through what he wanted to tell his staff, Landon's map began to fill in and become more symmetric. After a couple of passes, he was satisfied with the organization and was ready to produce a formal outline for his talk.

Example of a Mind Map

Figure 3.3 shows one possible mind map for a quick tutorial presentation on antenna measurements.

This mind map, which took only 5 minutes to generate, could obviously

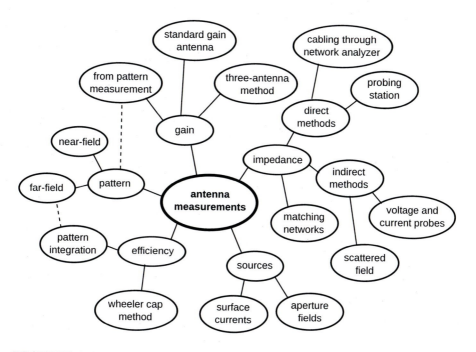

FIGURE 3.3

First author's mind map for antenna measurements.

serve as a basis for construction of a more detailed conventional outline, and then a set of projection slides. The outline might look like this:

Outline for a Seminar on Antenna Measurements

I. Coverpage

II. Background

    A. Why do we make antenna measurements?

        i. Communications systems

        ii. Radar

        iii. RFID

    B. Types of antennas

        i. Wire antennas

        ii. Aperture antennas

III. Equipment you will need

    A. Chamber

    B. Network analyzer

    C. Probes/antennas

    D. Ancillary equipment

IV. Types of measurements you can do

    A. Pattern

        i. Near field

        ii. Far field

    B. Gain

        i. Using standard gain antennas

        ii. Three-antenna method

        iii. Gain pattern

    C. Impedance

        i. Direct methods

            a. Cabled to network analyzer (VNA)

            b. Probe station

        ii. Indirect methods

            a. Voltage and current probes

            b. Scattered field

        iii. With matching networks

    D. Efficiency

        i. Wheeler cap method

        ii. Pattern integration

        E. Sources
           i. Surface current
           ii. Aperture field
    V. Example measurements
        A. Patch antenna
           i. Pattern
           ii. Gain
           iii. Impedance
           iv. Efficiency
        B. Dipole antenna (current)
        C. Horn antenna (aperture field)
  VI. Resources for further learning
        A. Classic articles
        B. Books
        C. Websites
 VII. Conclusion
VIII. Thanks!

Stuck? Some Approaches

If mind mapping fails to produce material for a suitable outline, you could try the following before making another attempt.

1. Sleep on it. Set the task aside and return to it the next day. (This highlights one of the advantages in attacking a project early; procrastinators often lose the opportunity to *sleep on it*.)

2. Learn more about the topic. Closing certain gaps in your knowledge might help you get going.

3. Look at examples of other presentations. Review the slides for successful talks you gave in the past, or search the web for inspiration.

4. Ask a colleague or expert for ideas. Someone who knows the topic well may be able to help you structure a talk appropriate for your purposes. Or just talk with them about your topic and take notes on where the conversation goes.

5. Ask a potential audience member for ideas. Someone in or close to the target audience might be able to assist by pointing out things that could be helpful or interesting for them to hear.

Finally, if none of these approaches gets you unstuck, then *just start with something*. Remember that the engineering design process is largely based on iterative improvement! Even a primitive initial configuration can evolve into a wonderfully differentiated talk when subjected to several cycles of recursive enhancement.

> **Example.** Emma had to pick a topic for an oral presentation in her Survey of Environmental Engineering course. It was nice to be given the latitude to select her own topic; on the other hand, Emma really didn't know what to choose or how far to go with it. Ron, another student in the class, listened to Emma's frustration and said, "I'm preparing my talk on the hydrogeology of waste disposal, but I've always wondered about the legal aspects of things like direct soil contamination. Who pays for all the cleanup, and why?" Here, then, was a topic that would interest at least one of her classmates. By chatting with a few more friends, Emma was able to identify a cluster of legal questions relevant to environmental engineering practice.

3.6   Checklist: Designing Your Presentation

☐ **I have established the structure of my presentation**

☐ **I have designed my introduction**

☐   I understand what I want to accomplish with my introduction

☐   I have decided on a structure for my introduction

☐ **I have designed the main body of my presentation**

☐   I know the type of presentation I will be doing

☐   I have decided on an order of presentation

☐   I have considered how I will persuade the listener

☐   *I have a way to gain the listener's attention*

☐   *I address the need for my product*

☐   *I offer a solution*

☐   *I differentiate my product from that of the competitors*

☐   I have created an outline of my talk, possibly using a mind map as a guide

☐ **I have designed the conclusion of my presentation**

☐   I understand what I want to accomplish with my conclusion

☐   I have an indicator of the end of the presentation

3.7 Chapter Recap

1. Every presentation needs an introduction, a main body, and a conclusion. The introduction and conclusion can be as brief as one sentence, but they have to be present.

2. Aim for smooth transitions in your talk, unless you wish to shock the listener (a risky practice at best).

3. An introduction or conclusion can take a variety of forms and accomplish a variety of things. In general, both function as smooth and clear transitioning elements.

4. There are many standard types of technical presentations (briefings, proposals, etc).

5. Classically, speeches tend to fall into certain general types (informative, persuasive, etc). The primary purpose of a technical presentation, as considered in this book, is to inform.

6. Engineers are often called upon to persuade. People are busy; they may not give your idea or proposal a chance if not provided with an adequate reason.

7. Mind maps are useful for organizing your thoughts. The information on a mind map can be converted to an outline, and subsequently into projection slides for use in your talk.

8. If you don't know where to start, then *start somewhere*; that's what iterative improvement is for.

3.8 Exercises

3.1. Todd started his presentation by saying

I'll be talking about AWGN today.

Comment.

3.2. Analyze the introduction and conclusion phases of a public lecture or TED talk.

3.3. Pick a topic, and formulate an introductory paragraph to accomplish each of the following things:

(a) getting the audience's attention,

(b) introducing your subject,

(c) giving the audience a reason to listen,

(d) establishing your credibility as a presenter,

(e) previewing your main ideas.

3.4. Pick a topic, and formulate one-paragraph introductions taking each of the following forms:

(a) introduction by reference to the occasion;

(b) introduction by reference to a recent incident;

(c) introduction by reference to the preceding speaker's remarks;

(d) introduction by reference to the subject;

(e) introduction by statement of your specific purpose or thesis;

(f) introduction by preview of points.

3.5. Pick a topic, and formulate a conclusion to accomplish each of the following things:

(a) summarizing what you said,

(b) stressing the main idea one last time,

(c) getting the audience to take some action after your talk,

(d) giving the listeners a sense of closure.

3.6. On page 32 we cautioned about the use of humor to open a technical presentation. For the presenter who is determined to incorporate humor, however, can you propose any sensible guidelines?

3.7. Construct a mind map for an interesting technical topic.

3.8. Generate a detailed outline for an interesting technical topic.

3.9. Construct a rubric to evaluate how well you have designed your presentation. You may wish to use the checklist from Section 3.6 as a guide.

4

Building Your Presentation

What considerations should go into producing the *content* of your presentation? Although the specifics are dependent on your topic, several general considerations are critical.

4.1 The Target Specifications

As engineers, we can't design a product without *specifications*. Similarly, as technical speakers we can't design a presentation without a clear picture of the specifications we're trying to meet. One sensible breakdown is as follows.

1. *The subject material.* What is the scope of your presentation? Which topics do you wish to cover (or need to cover), and at what depth?

2. *The audience.* To whom will you be speaking? What are their backgrounds and purposes?

3. *The time length.* How many minutes or hours will you have? Will the limit be strictly enforced? If the presentation lasts for hours, when will breaks be taken?

4. *The venue.* In what environment will you be speaking?

   (a) Indoors? Outdoors?

   (b) What are the acoustics and lighting situations like?

   (c) How will the audience be seated?

   (d) What sorts of equipment will be available?

Now is the time to answer as many of these questions as possible. The issues we have listed can vary markedly from one situation to the next. *Don't make the mistake of assuming that the variables pertaining to your last talk will carry over to your next one.*

Example. Larry was told that he would be giving a review of his recent project to one of the vice presidents. "OK, know your audience," he thought to himself. "The top brass don't like a lot of detail, so I'll just concentrate on the big picture." Then, he joked, "And use small words!" Larry couldn't have been more wrong. It turns out that this VP is a former engineer, and in fact had helped design the original product that Larry was now working to improve. Larry's presentation was a disaster. Afterwards he talked with a colleague who said, "This VP demands a lot of technical content. He wants to make sure you have the capabilities to handle the task. I wish you had talked with me first." Larry learned that *know your audience* isn't just a platitude, and that a little research can prevent painful misunderstandings.

4.2 Quality Control: Some Key Aspects

Good technical speakers put forth arguments that are logically sound. Their English is vivid, accurate, and appropriate; their math is clear and correct. They come equipped with effective, attractive visual aids. And they always bear in mind the ethics and potential legalities of a situation.

Example. Sara gave a stunning technical talk about automotive communication systems, but failed to disclose her relationship as a paid consultant for one of the major car manufacturers. Later she was accused of a breach of ethics for harboring a conflict of interest.

Your presentation will only be as strong as its weakest link. In the next few sections we provide some pointers about the diverse aspects of presentation quality.

4.3 Logic

Although logic is an extensive and beautiful subject worthy of study for its own sake, our interest is to prevent silly (or catastrophic) *logical blunders*. In speech as in writing, people commit two main types of logical mistakes or *fallacies*: formal and informal. Let's examine each of these in turn so you can avoid them.

Formal Deductive Fallacies

Formal fallacies violate basic, accepted patterns of formal reasoning. One such pattern is the classic categorical syllogism illustrated by the argument

> All men are mortal.
> Socrates is a man.
> Therefore, Socrates is mortal.

The first two statements are the *premises* (or *assumptions*) and the third is the *conclusion*. The argument is valid not because the conclusion holds, but because the truth of the conclusion must follow from that of the two premises. Any argument of the general form

> All S are P.
> x is an S.   ✓ valid
> Therefore, x is P.

is valid for the same reason. In Figure 4.1 we use diagrams to check this. The same syllogism may appear in abbreviated fashion:

> All men are mortal, so Socrates is mortal.
> Socrates is a man, so he is mortal.

In each case one of the formal premises is left implied.

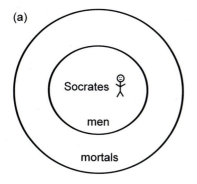
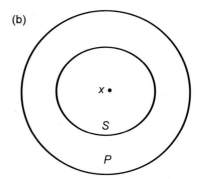

FIGURE 4.1

A categorical syllogism. (a) Concrete case. Socrates is a man, and all men are mortal. So Socrates is mortal. (b) Abstract case. The point x falls within the set S, which in turn lies within a set P. Clearly x must fall within P.

Example. While giving a talk on chemical safety, Caroline stated:

> Arsine is a Cat 1 toxin. It can kill you.

This argument takes the form:

> All Cat 1 toxins are substances that can kill you. (All S are P.)
> Arsine is a Cat 1 toxin. (x is an S.)
> So Arsine is a substance that can kill you. (Therefore, x is P.)

Since the premises are true, and since the argument is a valid syllogism, the argument is sound. As you can imagine, it had a powerful effect on Caroline's audience.

Now let's examine a couple of fallacious arguments. Consider

> All X is Y.
> y is Y.   × invalid
> Therefore, y is X.

and examine Figure 4.2(a). This argument form is *not* valid: the point y shown in the figure is an *invalidating counterexample*, and it takes only one of these to wreck the argument. The argument form

> All Y is X.
> No Y is Z.   × invalid
> Therefore, no Z is X.

is invalidated by the counterexample x in Figure 4.2(b).

Example. While giving his thesis presentation on microwave measurements, Joseph said:

> There are many types of RF connectors, but all are precision connectors. This is because 3.5 mm connectors are precision connectors, and some RF connectors are 3.5 mm connectors.

Several audience members winced at his conclusion that all RF connectors are precision connectors. They knew that it was not true, regardless of his

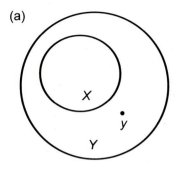 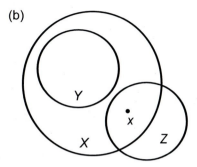

FIGURE 4.2
Two invalid syllogisms. (a) Everything that has the attribute X also has the attribute Y. However, the additional assumption that y has the attribute Y does not imply that it *must* have the attribute X. (b) The assumption that Z and Y are disjoint does not imply that Z and X are disjoint. An element x can belong to both Z and X.

contorted logical argument. A little careful analysis would lead Joseph to the underlying form of his argument (see Figure 4.3):

> All 3.5 mm connectors are precision connectors. (All T is P.)
> Some RF connectors are 3.5 mm connectors. (Some R is T.)
> So all RF connectors are precision connectors. (So all R is P.)

Both premises are true, so what went wrong? This is another example of an invalid categorical syllogism, so Joseph's *argument* is invalid.

In many ways, speakers must be quicker in their thinking than writers. As they don't have the luxury to stop and ponder over their statements, they must be adept at quickly analyzing the logic of their own arguments. A sturdy grasp of logical argument will help you avoid many embarrassing blunders. For convenience, we have provided a list of valid categorical syllogisms in Exercise 4.6.

Arguments Based on Conditional Statements

Not all reasoning patterns are based on categorical syllogisms, so we must cover a few more ideas. From now on, unless otherwise stated, uppercase letters such as P and Q will denote *statements* rather than classes of objects. So P could be

> the area of a circle having radius r equals πr^2

which is a true statement, and Q could be

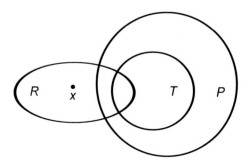

FIGURE 4.3
Joseph's logical blunder. T lies completely within P, and *some* of R lies within T. But it does not follow that *all* of R *must* lie within P, as demonstrated by counterexample x.

-14.8 is a positive integer

which is a false statement. A compound statement of the form

If P, then Q.

is a *conditional statement*; the statement P is its *antecedent* and Q is its *consequent*. Take a moment to memorize these terms; they are standard and will occur repeatedly in the next few pages.

Example. Consider the conditional statement:

If I am a licensed professional engineer, then I have a college degree.

Let's break it down:

If $\underbrace{\text{I am a licensed professional engineer}}_{\text{antecedent, } P}$, then $\underbrace{\text{I have a college degree}}_{\text{consequent, } Q}$.

Note that we are not free to interchange the antecedent and consequent of a conditional statement without a (possibly drastic) change in meaning.

Example. The statement:

If I have a college degree, then I am a licensed professional engineer.

is clearly false; as valuable as a nursing degree may be, it does not grant licensure as a professional engineer. This statement is called the *converse* of the statement in the preceding example.

We now examine some argument forms involving conditional statements. Please don't be discouraged by their technical sounding names. Arguments of these forms are recognizable in all engineering discourse.

Modus Ponens, or Affirming the Antecedent

The pattern

If P, then Q.
P.   ✓ valid
Therefore, Q.

is a standard argument form called *modus ponens*. Since the antecedent of the conditional in the first premise is affirmed by the second premise, the form is also called *affirming the antecedent*.

Example. The argument:

If my presentation is bad, I won't get the contract. My presentation is bad. So I won't get the contract.

takes the form of *modus ponens*. It is a valid argument.

Modus Tollens, or Denying the Consequent

The following pattern is a standard argument form called *modus tollens*. By *not-Q*, we mean the statement called the *negation of Q*.

If P, then Q.
Not-Q.   ✓ valid
Therefore, not-P.

In English we can negate a statement by appending *It is false that* to the start of it, although this may not yield the most concise or graceful formulation.

Example. The statement:

All keynote speakers are college professors.

is false. Its negation can be phrased in any of the following ways:

It is false that all keynote speakers are college professors.

Not all keynote speakers are college professors.

Some keynote speakers are not college professors.

There is at least one keynote speaker who is not a college professor.

The negation of a false statement is true, and the negation of a true statement is false. Let's get back to *modus tollens*. Since the consequent of the conditional in the first premise is denied by the second premise, the form is also called *denying the consequent*.

Example. The argument:

If my presentation exceeded 30 minutes I would have been stopped by the moderator. I was not stopped by the moderator. Therefore my presentation did not exceed 30 minutes.

takes the form of *modus tollens*. It is a valid argument.

An Argument Form with Two Conditional Premises

Here's another standard reasoning pattern, this time with conditional statements for both premises.

If P, then Q.
If Q, then R.   ✓ valid
Therefore, if P then R.

Example.

If my abstract exceeds one page, it is rejected. If my abstract is rejected, I cannot present my paper at the conference. So, if my abstract exceeds one page, I cannot present my paper at the conference.

Fallacies Involving Conditional Statements

What can go wrong with arguments containing conditional statements? There are two famous fallacies to guard against. The first is called *affirming the consequent*.

Example. The argument:

> If my abstract is not properly formatted, it is rejected. My abstract is rejected. Therefore my abstract is not properly formatted.

is not *modus ponens*. The antecedent of the conditional ("my abstract is not properly formatted") is not affirmed in the second premise; rather, the consequent ("it is rejected") is affirmed. This argument is invalid. (There may be many reasons for rejecting an abstract other than formatting.)

In general, an argument of the following form is fallacious.

If P, then Q.
Q.   × invalid
Therefore, P.

Again, this is called affirming the consequent. Let's proceed to the second fallacious form.

Example. The argument:

> If this is Tuesday, we have a review meeting. This is not Tuesday. Therefore we do not have a review meeting.

is not *modus tollens*. The consequent of the conditional ("we have a review meeting") is not denied; rather, the antecedent ("this is Tuesday") is denied. This argument is invalid. (Maybe review meetings occur on both Tuesdays and Fridays.)

In general, an argument of the following form is fallacious.

If P, then Q.
Not-P.   × invalid
Therefore, not-Q.

This is called *denying the antecedent.*

The Disjunctive Syllogism

Another valid argument form, commonly seen, is the *disjunctive syllogism.* The pattern is

P or Q.
Not-P.   ✓ valid
Therefore, Q.

The first premise guarantees that at least one of the statements P and Q must hold. This, taken together with the second premise (that P does not hold), is enough to guarantee that Q holds.

Example. The argument:

> Either the presentation is canceled or I am in the wrong room. The presentation is not canceled. Therefore I am in the wrong room.

takes the form of a disjunctive syllogism. It is valid.

Informal Fallacies

The fallacies presented above are examples of formal, deductive fallacies. Another part of logic, called *informal logic*, collects and classifies other types of fallacies that commonly occur in human discourse. People have committed these types of errors intentionally or unintentionally since the time of Aristotle (384 – 322 BCE). Avoid them all.

Ad Hominem

This fallacy occurs when someone attacks a person rather than his argument.

Example. Henry is giving a presentation to several managers and engineers to convince them to start using a new process he has developed.

> I know Dan argues that this new process is less efficient than the current one, but you know Dan. He's never willing to try anything new. He's always stuck in the past.

It may be that Dan is excessively cautious, but that does not discredit his arguments regarding the new process.

Fallacy of Accident

This fallacy occurs when someone applies a rule to a case it was not intended to cover.

Example. Previously we told the story of Larry, who had heard that the top brass only want the "big picture" and that one should never bother them with details. Larry fell victim to the fallacy of accident when he applied this rule to a VP who himself was an engineer. His logic went something like this: (1) bigwigs don't want to hear about technical details; (2) the VP is a bigwig; (3) he won't want to hear about details. Larry was caught applying a rule of thumb (bigwigs don't like details) where it isn't appropriate — in this case thinking a bigwig who is an engineer won't want to hear technical details. Sometimes called a "sweeping generalization," this fallacy can cause us headaches when we apply rules of thumb to situations where they are not valid. Engineers are particularly vulnerable when they learn about rules without knowing their limitations.

Straw Man Fallacy

This fallacy occurs when someone distorts a person's position and then attacks the distorted version.

Example. Allen was presenting his ideas to his manager about a new annealing process he had devised to strengthen beryllium alloys. His colleague, Mannie, had proposed instead that the company should use Klingman's process, a phase transition approach accepted industry-wide. Allen knew that to get his manager to buy into his idea he would have to provide a good reason, so he decided to attack Mannie's position.

> Everyone knows that phase-transition processes are just too expensive! Look at Johnson's process — no one uses that anymore because it costs too much.

His manager jumped in:

> Wait a minute. Mannie never said he wanted to use Johnson's process. It has nothing to do with annealing!

Allen had used Johnson's process as a straw man to attack Mannie's idea; unfortunately for Allen, his manager caught on right away.

Appeal to Ignorance

This fallacy occurs when someone gives up on further thinking and investigation. They might say, for instance, that event A must have caused event B because they cannot imagine any other reason for the occurrence of B.

Example. In Clara's first conference presentation, she examined the proposition (suggested by some in industry) that coating a waveguide tube with carbon can prevent unwanted microwave oscillations. Her theoretical analysis showed that this approach generates too much loss, and her last slide asserted that:

> Coating the tube with carbon is not a viable option for eliminating oscillations.

The very next speaker started his talk with:

> The last speaker gave a great argument of why carbon can't be used to suppress oscillations. Now I'll show you how we did it.

He went on to show how, by slightly altering the pattern of carbon, the loss could be reduced significantly and the oscillations eliminated. Clara was mortified and has never forgotten the lesson. You can't assume a proposition is false just because no one has proven it true (or vice versa).

Hasty Generalization

This fallacy occurs when someone makes an inference about all members of a group from the characteristics of an insufficient sample.

Example. Tim is a graduate student in chemical engineering who has now given two conference presentations. Both presentations were at the same conference. His colleague Penny was to give her first talk at a conference that, although similar in theme, was different than the one Tim attended. Tim said:

> The best thing about these conferences is that you can use your own computer to do the presentation.

Penny brought her computer, but was horrified to discover that presenters were required to use a specific system she had not prepared for. Fortunately she was able to get emergency help in adapting her presentation materials for the new system. Tim lacked sufficient experience with conference presentations to make a general statement about the procedures at all conferences. Be sure to do your homework!

Post Hoc Ergo Propter Hoc

This fallacy occurs when someone concludes that event A must have caused event B because A preceded B in time.

Example. In his master's defense, Isaac was faced with the need to explain why some of his measurements were considerably worse than others. He had noticed that all of the "bad" measurements were made after the installation of a new balance.

> I'm pretty sure that the cause of my problems is an improperly calibrated balance. It can't just be a coincidence that all my troubles started after it was installed.

In fact, it *was* just a coincidence, and his adviser let him know it:

> Wait a minute! I had that balance calibrated right after installation. We have since used it in many measurements, and all have turned out well. Think again!

Isaac should have given this issue more thought before his presentation. Just because the balance was installed before his measurements went bad, doesn't mean it was the cause of his troubles.

Cum Hoc Ergo Propter Hoc

This fallacy occurs when someone concludes that event A must have caused event B because A and B occurred simultaneously.

Example. Bill really put his foot in his mouth when he made this statement during a briefing on his project:

> Our funding got cut when the Division budget got reduced, and this is making it really hard to complete the project.

His boss jumped in:

> You didn't get your funding cut because of this! Rather, your progress was too slow and so we decided to cancel your project. Had you done well, I would have protected you from the budget reductions!

Ouch.

Fallacy of Composition

This fallacy occurs when someone erroneously attributes a trait possessed by all members of a class to the class itself.

Example. Aubry fell victim to the fallacy of composition when she gave a talk about sintering at a regional meeting of professional engineers. During the rehearsal for her talk, she received feedback that she should slow down and provide additional detail when describing the flowchart for the sintering process. This worked so well that Aubry hastily decided (during the presentation) that what's good for one slide should be great if applied to every slide. Unfortunately she failed to consider the aggregate increase in time that changing her pace would incur, and she exhausted her allotted time before reaching the halfway point in her talk. While it probably would help on any one slide to slow down a bit, slowing her pace on the whole talk was a disaster.

Fallacy of Division

This fallacy occurs when someone erroneously attributes the traits of a class of objects to each of the separate objects.

Example. While presenting her senior project, Donna claims that the dipole antenna used in her communications link has a high gain. She is quickly corrected by her instructor, who notes that a dipole has a relatively small gain.

> But the claim of the manufacturer I read said that dipoles have large gain.

After much back and forth, the instructor finally teases out that Donna has used one element from a large antenna array. While the array itself has high gain, each element has relatively low gain. The gain is high only when the elements act together.

Begging the Question

This fallacy occurs when someone uses their target conclusion as one of their premises.

Example. Carlos made the mistake of begging the question during his dissertation defense, when he made the statement:

> This chemical is highly acidic because it has a low pH.

This led to a lengthy discussion of whether Carlos really understood the chemical he was studying. In essence, Carlos made the argument that "the chemical is acidic because it is acidic," and this was not very convincing to his examining committee.

Weak Analogy

This fallacy occurs when someone argues based on an alleged similarity between two situations that, in reality, are not that similar.

Example. Analogies are a great way for a speaker to connect with her audience, but care is required to make sure that such analogies do not lead to incorrect conclusions. For instance, to help the audience visualize ductility, Belinda could say during a talk on the strength of materials that:

Gold stretches when pulled, just as rubber does.

But for her to conclude that "gold and rubber must have similar underlying structures" would be silly.

False Dichotomy

This fallacy occurs when someone bases an argument on the premise that either A or B must hold, when in reality a third possibility C could hold.

Example. During her project presentation in a mechanical engineering class, Pam's instructor tried to fluster her by trapping her in a false dichotomy.

Is the mass in a state of stable or unstable equilibrium?

She refused to be nonplussed and responded:

It's not in equilibrium at all. The mass is accelerating.

Pam avoided being trapped in a fallacy.

Fallacy of Suppressed Evidence

This fallacy occurs when someone omits counterinstances while drawing an inductive conclusion.

Example. Rob presented the results from his study of automobile accidents to a safety review panel. The issue was whether the failure of a part manufactured by his company was at fault in a string of collisions. He quoted heavily from research done at a local university, and at the end of his talk summarized his findings.

The evidence does not support the hypothesis that the failure of the coupling caused the accidents under investigation.

One panelist was not satisfied.

> You completely overlooked two studies done by the National Safety Institute. I think when you consider their findings, your conclusion will be quite difficult.

Scientists and engineers must always refrain from "cherry picking" their results. If you don't trust certain results, don't dismiss them, but instead explain to the audience why you find the results questionable.

Fallacy of Equivocation

This fallacy occurs when someone uses a word in two different ways in the same argument.

> **Example.** Jenny had an embarrassing moment during her project presentation for her class on fermentation when she was asked why she didn't purge her system. Wasn't she worried about Henry's law?
>
> > We didn't think it applied to us because we checked the local statutes and couldn't find it listed anywhere.
>
> Jenny had confused "scientific law" with "legal regulation."

Fallacy of Amphiboly

This fallacy occurs when someone argues based on a faulty interpretation of an ambiguous statement.

> **Example.** It is easy to perplex an audience with poorly constructed statements. Consider the audience's confusion over George's statement:
>
> > I decided to do the error analysis after I looked at the results.
>
> Did he mean
>
> 1. "After I looked at the results, I decided I should do the error analysis," or
>
> 2. "I decided that I should look at the results, and then do the error analysis."
>
> The difference could be significant.

Appeal to the Crowd

This fallacy occurs when someone argues that statement A must be true because most people believe it's true.

Example. During Annette's thesis defense, she was asked why she calibrated her system at the end of the experiment and not at the start. Her reply was:

> Everybody in our group has been doing it that way forever. They can't all be wrong.

Yes, they can. It turns out that once one person had calibrated the system improperly, everyone else followed suit.

Fallacy of Opposition

This fallacy occurs when someone argues that statement A must be false because their opponent believes it's true.

Example. Martin assessed a potential new process in a briefing to his boss:

> If those guys at Cromwell Machines are using the process, then I wouldn't put too much faith in it. We can always do better than them. Let's use another process.

It may turn out that this is the best process to use, and Martin shouldn't be so quick to dismiss it without solid evidence.

The fallacy of opposition often leads to missed opportunities. Factions form within companies, within universities, within research groups. Just because someone is your competitor, doesn't mean you are right and they are wrong.

Appeal to Authority

This fallacy occurs when someone argues that statement A must be true because experts believe it's true.

Example. There's no doubt that quoting an expert during your presentation can lend credence to your argument, but don't forget that even the "experts" can be wrong. Consider what the great physicist Niels Bohr once said:

An expert is a person who has found out by his own painful experience all the mistakes that one can make in a very narrow field.[1]

Be sure you have corroborating evidence.

Additional Ways to Check for Errors

The following approaches should be familiar to any engineer.

Intuitive Plausibility

Is the result reasonable? This question can save us much embarrassment in
front of an audience.

Example. While generating numerical data for a presentation, we notice
that when the temperature in our model is reduced, the thermal radiation
is increased. Perhaps we should look for an error in our calculations.

Dimensional Checks

Physical dimensions must match across all terms in a valid physical equation.

Example. Suppose we write:

The maximum height of a projectile launched from an initial height
h_0 at an angle θ with velocity v is given by

$$h = h_0 + \frac{v^2 \sin^2 \theta}{g} \tag{1}$$

where g is the acceleration due to gravity.

The physical dimensions in (1) are analyzed as

$$[\text{length}] = [\text{length}] + \frac{[\text{length}]^2}{[\text{time}]^2} \Big/ \frac{[\text{length}]}{[\text{time}]^2}$$

and thus both terms have the same dimensions of [length]. This is *necessary* for the equation to be correct, but it is not *sufficient*. It doesn't
help us learn whether there is a missing term, or if the power on the sine
function is correct (since the sine function is dimensionless). Even so, a
dimensional check is a valuable method for finding errors.

---

[1] "Dr. Edward Teller's Magnificent Obsession," *Life Magazine*, p. 62, September 6, 1954.

Example. If x and y have units of length, then the equation

$$x = \cos y$$

is wrong for two reasons. The cosine function is dimensionless and so the dimensions cannot match across the equals sign. Additionally, the cosine function is required to have a dimensionless argument.

More practice with dimensional checks is provided in Exercise 4.7.

Order-of-Magnitude Checks

Quantitative claims should be numerically reasonable.

Example. Marie's senior project group is preparing for their final presentation. She has been assigned the task of estimating the distance that a potato will travel when launched by a pneumatic potato gun. She plugs numbers into the equation

$$d = \frac{v^2 \sin(2\theta)}{g}$$

and obtains the answer 1×10^{13} m. This seems a bit large to her, being greater than the distance from the earth to the sun. A quick check reveals that she has used "big G" (the universal gravitational constant) instead of "little g" (the free-fall acceleration constant) in her calculation. Recomputing with the correct constant, she finds $d = 70$ m, a much more reasonable result.

Expected Variation with a Parameter of the Problem

Some quantities in a problem may be denoted by letters but *temporarily held fixed during a calculation involving other variables*. We call these quantities *parameters*.

Example. The equipotential surfaces for a small electric dipole centered at the origin and aligned along the z-axis are given by the equation

$$r = c_v \sqrt{\cos \theta}.$$

Here r is the distance from the origin to the surface at an angle θ measured from the z-axis. This equation describes a *family* of surfaces: one for each value of c_v. To plot one equipotential surface from the family, we might set $c_v = 2$ and plot $r = 2\sqrt{\cos \theta}$. To plot another surface, we could set $c_v = 3$.

While plotting each surface of the family, we are holding c_v constant. However, c_v is neither a "true" constant like e or π nor a variable like θ; it is an example of a parameter.

Often an answer may be checked by varying parameters and validating against known behavior.

Example. Consider a simple problem from introductory physics. A particle carrying charge Q is fixed at the coordinate origin, while a second particle of charge q is moved from radius b to radius a in its presence. If $b > a$ then the work required to move the second particle is given by the integral

$$W = -\int_a^b \vec{F} \cdot d\vec{\ell} = -\int_a^b \frac{1}{4\pi\epsilon_0} \frac{qQ}{r^2}\,dr = \frac{qQ}{4\pi\epsilon_0}\left(\frac{1}{a} - \frac{1}{b}\right).$$

We can check this answer in several ways. If we increase either q or Q, the amount of work required increases since the force between the particles increases. This makes sense. If we increase b, the work required increases because we have to move the second particle farther. Again, this makes sense. Finally, if we decrease a, we again need more work to move the second particle farther.

As engineers, we should always be thinking this way. Using logical reasoning, we would easily catch an integration error leading to

$$W = \frac{qQ}{4\pi\epsilon_0}\frac{1}{\sqrt{ab}}$$

even though the dimensions are correct!

The above example shows the value of solving problems using parameters. We could have considered a fixed 5 C charge, and moved a 3 C particle from a radius of 2 m to a radius of 1 m, thereby considering just one specific case. Instead, by using parameters, we obtained a formula giving the work required to move between any two points for any two charges.

Agreement with Known Special Cases

One advantage of working problems in terms of parameters is that a problem may have limiting cases whose answers are known.

Example. Say that you are required to find the stopping distance of an automobile when the brakes are locked while the vehicle is traveling uphill.

You easily find the formula in a handbook for the case when the vehicle is braking on a flat surface:

$$D = \frac{v_0^2}{2g\mu_s}$$

where v_0 is the initial speed, g is the acceleration due to gravity, and μ_s is the coefficient of friction for the tires against the road. But, try as you may, you can't find the formula for a car traveling uphill. So you oil up your rusty calculus skills and derive the following formula:

$$D = \frac{v_0^2}{2g(\mu_s + \tan\theta)}$$

where θ is the incline angle. Is this correct?

First you check the dimensions. Since both μ_s and $\tan\theta$ are dimensionless, adding them is allowed and the dimensions check out OK. Next you decide to use some physical reasoning. If you increase the angle θ the car should decelerate and stop in a shorter distance. Sure enough, increasing $\tan\theta$ decreases D. Finally, you check against the known special case and let θ go to zero for a flat incline. Since $\tan\theta$ also goes to zero, your formula does reduce to the special case you found earlier. This gives you a lot of confidence that your answer is correct, but there is no guarantee that you did not make an error. For instance, the formula

$$D = \frac{v_0^2}{2g(\mu_s + \tan^2\theta)}$$

passes all three of the same tests. Nevertheless, checking for agreement with known special cases is always a good idea. It represents another tool you can use to hunt for errors in claims before you show them to the world during your next technical presentation.

You can also run checks by approximating an answer: dropping small terms, ignoring slow time variations, etc. These and many other useful techniques receive extensive coverage in the standard engineering curricula.

Other Mathematical Properties of the Answer

If an answer is time dependent, you might check its initial value, its final value, or its time-average value.

Example. Suppose your answer for the voltage in a circuit is

$$v(t) = 5 + 10e^{-5t}\cos 20t \text{ V}.$$

The initial value of $v(t)$ is $v(0) = 15$ V, and the final value is $v(\infty) = 5$ V. If either of these seems wrong, you should check your calculations.

Mathematical answers sometimes contain features such as infinite singularities and jump discontinuities. These are often (but not always) inappropriate from a physical standpoint.

Example. The displacement of a drum head in natural oscillation can be found by solving the wave equation. Proceeding along purely mathematical lines, the displacement is found to be proportional to the term

$$AJ_m(\lambda_{mn}r) + BN_m(\lambda_{mn}r).$$

Here r is the distance from the center of the drum head, λ_{mn} is the vibrational parameter for the (m, n) mode, and $J_m(x)$ and $N_m(x)$ are the Bessel and Neumann functions. However, from physical reasoning it is inappropriate to retain the Neumann function, even though it is a proper mathematical solution. Since $N_m(x)$ becomes infinite as $x \to 0$, this term must be discarded. After all, it makes no sense for the displacement of the drum head to become infinitely big at the center!

Accord with Standard Physical Principles

Notions like causality and symmetry are encountered routinely in the physical and engineering sciences. Why not use them to look for errors whenever possible?

Example. An electric circuit is excited for the first time at time $t = 0$, and you calculate the voltage as

$$v(t) = V_0 e^{-10|t|} .$$

This result looks promising since the voltage decays to zero as time increases to infinity. However, you quickly realize that the voltage exists for time $t < 0$, which is *before* the excitation was applied. Because this violates the principle of causality, your formula cannot be correct.

Another important principle is superposition, although one must remember that it applies only to linear systems.

Example. Always check to see whether the whole is greater than the sum of its parts. When adding sinusoidal signals, for instance, the amplitude of the resulting sinusoid can be smaller than the sum of the individual amplitudes, but it can never be larger. Similarly, in linear media, two strong waves may come together and cancel each other by destructive interference, but they may never add to a wave ten times as big as the larger of the two.

Other Ways to Be Careful

Follow these suggestions to avoid sloppiness and prevent errant claims.

- Don't jump to conclusions.

- Maintain a critical attitude.

- Respect the truth.

- Employ Ockham's razor.

- Insist on reliable evidence from dependable sources.

- Double check everything.

- Always seek counterexamples to your claims.

See the authors' book *Engineering Writing by Design: Creating Formal Documents of Lasting Value* for expanded treatment of the points in this list.

Example. The *post hoc ergo propter hoc* and *cum hoc ergo propter hoc* fallacies can rear their ugly heads when inappropriately linking cause and effect. Remember that *correlation does not imply causation.* The fact that your team won the big game when you were wearing your shirt backwards is no reason for you to do the same during every game. As an engineer, you probably wouldn't fall victim to such a silly superstition (would you?). However, if you notice that those pesky oscillations disappeared when you lowered the manifold temperature, you should probably run a sequence of controlled experiments before deciding that there is an actual relationship between the events.

4.4 English Usage

Crucial among the building blocks of any technical presentation are the *words* the speaker uses and how they are connected together.

Word Choice

Some words may appear on your slides and some may be "merely" spoken, but they are *all* important and worthy of careful selection. The right choice of words will take into account two issues:

1. correct technical meaning, and

2. proper level of formality for the occasion.

A technical presentation is not the time or place for sloppy diction. Informal terminology, slang, cute spelling, and internet jargon are not appropriate in a formal presentation, unless specifically relevant to the topic under consideration. Acronyms or specialized terms need to be written out or explained. Remember that clarity of presentation is key.

Example. Anna, a young engineer at a large electronics firm, was making her first presentation at a program review meeting. She was used to sending out quick, informal briefings to her teammates through e-mail, but now had to describe their recent progress to her manager and his staff. She gathered together her briefings and put them into a short presentation. One slide, an update on the progress of the computer coding, contained the following bullet points:

- Pam said to give her a hedzup B4 we start coding — she's going PLOA

- Joe said to do up the interface design first, and then handle the coding l8tr

- I suggested we aughta use Git for sharing code. This really rox!

- Next step — we hafta integrate across platforms, IMHO.

As you might expect, Anna's boss reacted to this lack of professionalism with a combination of confusion, embarrassment, and disappointment.

Correct spelling is a must. When in doubt, always consult a reliable dictionary.

Example. Beware of words that look alike or sound alike. Sometimes a spell checker isn't enough. Imagine having these on your slides:

We used aesthetic acid.

The break pads were warn.

It will consume an energy of 10 dines.

Water friezes at 0 degrees Centipede.

The you-bolts were the week lynx.

We use sigh to represent the angel in poler coordinates.

And, although this may seem obvious, be sure you can actually pronounce all the words on your slides!

Example. In the heat of a cut-and-paste session, Brett included the word *algorithm* on one of his presentation slides. Imagine everyone's discomfort when, during the actual event, Brett took several stabs at saying this word before giving up. He knew what an algorithm was. But different levels of familiarity with words are possible — the fact that you can *recognize* a word doesn't mean you can *say* it correctly ... especially with 50 people staring at you.

A list of commonly mispronounced words is given in the table below. The phonetic translations are ours, and use the 40 phonemes of the system *Truespel* (http://www.truespel.com/en/).

word	incorrect pronunciation	correct pronunciation
across	u-krausd	u-kraus
arctic	aar-tik	aark-tik
asterisk	as-ter-iks	as-ter-isk
athlete	athh-u-leet	athh-leet
cavalry	kal-ver-ee	ka-vool-ree
dilate	die-u-laet	die-laet
escape	ek-skaep	e-skaep
et cetera	eks-set-ru	et-set-er-u
February	feb-yue-wair-ee	feb-rue-air-ee
figure	fi-ger	fig-yer
foliage	foil-ij	foe-lee-ij
height	hiet-thh	hiet
jewelry	jue-ler-ee	juel-ree
liable	lie-bool	lie-yu-bool
library	lie-bair-ee	lie-brair-ee
nuclear	nue-kyue-ler	nue-klee-yer
ordnance	or-di-nints	ord-nints
picture	picher	pik-cher
prerogative	pir-aa-gu-tiv	pri-rraa-gu-tiv
prescription	per-skrrip-shin	pree-skrrip-shin
probably	praab-lee	praa-bub-lee
Realtor	reel-u-ter	reel-ter
supposedly	su-ppoez-ub-lee	su-ppoez-ed-lee

Be particularly careful when quoting someone else's words. Nothing will get you into hot water faster than quoting incorrectly or out of context, and then finding that the source of the quotation is in the audience! Be sure to vet the quote carefully by comparing it to the primary source, or by directly speaking with the person quoted, and avoid quotations that might lead to a "he said, she said" situation. And always verify the accuracy of any quotation in a foreign language.

Example. A BBC report tells the story of an official in Wales whose e-mail request for a translation of a road sign from English to Welsh led to a humorous situation.[2] The official did not speak Welsh and did not verify the validity of the received reply. As a result, the following statement was placed on the sign in Welsh: "I am not in the office at the moment. Send any work to be translated."

Punctuation

Proper punctuation is essential in formal technical documents, and we devote much space to this topic in our companion volume *Engineering Writing by Design* (see page 137 for the full reference). Our position regarding presentation slides is that punctuation should be handled consistently.

Example. In the absence of explicitly stated punctuation guidelines or requirements, we believe the following lists are all acceptable:

Magnets:	**Magnets**	**magnets:**	**magnets**
• Physics.	– physics	▶ physics	physics
• Materials.	– materials	▶ materials	materials
• Shapes.	– shapes	▶ shapes	shapes
• Uses.	– uses	▶ uses	uses

It is clear that many other acceptable formats exist. However, contrast with these examples:

Magnets:	**Magnets**	**magnets:**	**magnets**
• Physics	– physics,	▶ physics	– physics
• materials	– Materials,	▶ materials	– materials
• shapes	– shapes	▶ Shapes	– shapes
• Uses	– uses	▶ uses.	Uses

Note the various distracting inconsistencies.

---

[2]http://news.bbc.co.uk/2/hi/7702913.stm. Last viewed January 24, 2015.

Through random capitalization and inconsistent punctuation, you could disorient your listeners and convince them that you lack sufficient attention to detail.

> **Example.** Does Dmitri's slide describing how he obtained carbon dioxide give you much confidence in the outcome of his experiments?
>
> - the sherman johnson co inc produces the BEST SOURCE of CO_2 for our EXPERIMENTS.
> - We ordered 300 ML Bottles;
> - website/sjohnson.com/order-worked today
> - U can even get stuff on saturday!!!

Description versus Argumentation

We have said that technical presentation is a combination of information and persuasion. The process of informing typically entails description and argumentation. Presenters must describe ideas, objects, devices, or systems. They must also put forth logical arguments. It's important to understand that description and argumentation are not the same thing. Something you should be clear about, before opening your mouth to speak, is

> Am I trying to *describe* something right now, or *argue in favor of* (or possibly *argue against*) something?

Suppose, for example, you are presenting a mechanical system design. A full presentation may require both description and argumentation. However, you definitely want to be clear as to which of these you are pursuing at any given moment. You could, for instance,

1. start with a pure description of what the system *is*, then, after finishing the description, provide arguments for why the system should be (or had to be) that way, or

2. start with a set of standard design principles and show, through logical argument, how the system structure arises.

Either of these approaches is reasonable. It's *not* reasonable to oscillate between a description of something and a loose collection of arguments regarding why it must be the way it is. That will only confuse the listener.

Example. The following passage is clearly descriptive.

> In a rear-wheel drive automobile, the longitudinally-mounted engine transfers torque through the clutch to the transmission. The transmission connects to the drive shaft through a U-joint, and the drive shaft to the differential through another U-joint. The axle shafts then transfer the engine torque to the wheels through CV joints.

It tells the listener *what is*, not *why it had to be that way*. It is not argumentative. The speaker does not draw conclusions from given facts (deductive argument) or try to generalize from known facts (inductive argument). He or she concentrates, at least for the moment, on painting a picture.

Example. This passage is clearly argumentative:

> The point-to-point communication link implemented in our project requires the use of a high-gain antenna. Possibilities include reflector antennas, phased arrays of patches, and Yagi-Uda dipole arrays. We have chosen to use a 10-element Yagi-Uda array because of its low cost, simplicity of construction, and insensitivity to harsh environmental conditions.

Example. Consider this snippet of a talk:

> Let's take a look at our system. The input signal goes in the front. Sometimes you have to have two outputs for high efficiency. The output signal comes out here. Sometimes the signal comes out of the other side because of high efficiency. When it doesn't, we can't get our high efficiency. So we have two outputs. Remember, folks, it's all about efficiency!

Did you find this hard to follow? The passage is neither description nor logical argument, but merely an example of *bad* engineering discourse. *Don't do this.*

Cues for Logical Argument

You can tell that logical argument is taking place by looking for these key elements

1. *definitions* ("Suppose we let m be the mass of the ...")

2. *logical implications* ("if ... then ...") and

3. *logical equivalences* ("... if and only if ...").

Key words to look for are *premise indicators* such as *since, because,* and *in view of the fact that,* and *conclusion indicators* such as *therefore, hence,* and *we conclude that.* Learn to recognize and use these appropriately during your presentations, so that you don't sound like the speaker in the preceding example. For reference, let's list some common premise and conclusion indicators in English.

premise	conclusion
since, because	therefore, hence, thus
in view of the fact that	we conclude that
by virtue of	it follows that
for	consequently
inasmuch as	we may infer that
as indicated by	which implies that

Again, these should only appear in argumentation.

Some Pointers on Description

If you want to deliver a clear and informative description, pay close attention to the following classical recommendations.

(a) Keep your standpoint clear.

This is particularly important in engineering. Was your photograph of the construction site taken looking to the east, or to the west? Does it view upslope, or downslope? Was it taken in the spring during a period of temporary flooding, or during late summer so that the standing water is permanent? Each piece of information could be crucial to the audience's understanding of your explanation. If you switch between views as you scroll through your slides, be sure to keep the audience oriented. Make your descriptive standpoint clear.

Example A. Moira is describing her new technique for reducing corrosion inside electrical motor housings.

> On this slide you see five figures. At the center is a photograph of the interior of the motor housing, with a position marked with a red X. At the top are two micrographs made from samples at this position. On the left is a sample that was made before the treatment started. On the right is a sample taken five weeks after treatment. For comparison, the figures at the bottom show samples from the same two times taken from an untreated portion of the

housing. It is clear that the treated sample has experienced far less corrosion than the control sample.

Moira was careful to orient the audience to the physical position of the samples (inside the housing) and also to the temporal position of the measurements (five weeks apart). This gave the audience the information they needed to assess the validity of Moira's conclusion regarding the corrosion treatment.

(b) Put great thought into selection of details.

Include what is necessary and no more.

> **Example.** In Example A above, Moira could have said "We began the testing on August 24, and ended on September 30." This also provides the information that the test lasted five weeks, but that number must be teased out of the dates. Unless the time of year of the tests was important, specifying the duration was sufficient.

(c) Arrange the chosen details well.

Provide the listener with a coherent picture. If certain details are grouped naturally in your subject, they should be grouped intentionally in your presentation. Don't jump around haphazardly.

> **Example B.** Contrast the description of Example A with the following.
>
> > The treated sample turned out the best, as you can see. Both samples were taken after five weeks. The control sample is shown on the bottom and the treated sample is shown on top. The samples were taken from inside the housing. The left figures show before the treatment, and the right pictures show after. The control was not treated.
>
> Confusing, yes?

Crucial details can be emphasized by placing them first or last in the description. However, be careful that sufficient background information has been provided for the audience to understand the material described. In Example B, Moira emphasized the fact that the treated sample has less corrosion by mentioning this first. However, the audience had no point of reference by which to gauge the importance of the conclusion. In Example A, Moira waits until the audience is prepared, then makes a strong concluding statement based on the information she has provided.

These considerations show how important it can be to *have a plan* for delivering a description. Good descriptions don't just happen — they are deliberately and skillfully *designed*.

Other Aspects of English Usage

Ideally, a technical presenter will have all the grammatical skills of a good technical writer. This is not to imply that you must speak in the same formal way that you write; overly formal speech is generally unappealing, even in a technical presentation. However, this does not give you free rein to pepper your language with slang, cliches, or regional colloquialisms. You should strive to balance the level of formality needed to accurately convey your information with a genial tone that makes the audience feel comfortable.

We would make slightly different arguments about material *written* on your slides. Although we would not consider a slide as a formal document subject to the rules of formal grammar, we think it's best if slides do not say *blatantly ungrammatical* things.

Example. Imagine seeing the following bullet point on a technical slide:

- their are too solutions to equation (1)

First you'd have to translate it as

- *there* are *two* solutions to equation (1)

Then you might be annoyed by the fact that the presenter didn't write instead

- equation (1) has two solutions

which is shorter and more direct. But a mere confusion over English choices such as *their/there/they're* and *to/two/too* could cast doubt on the presenter's attention to detail.

There are simply too many of these issues to cover in a book on technical presentation. A reader interested in deepening his or her grasp of formal engineering writing could consult our companion volume *Engineering Writing by Design: Creating Formal Documents of Lasting Value*. Again, the full reference appears on page 137.

4.5 Mathematical Discourse

The art of communicating mathematical content is an extensive one encompassing both written and spoken phases. In this book, we assume you are an engineer preparing to address an audience composed mostly of engineers. Even so, depending on your area of specialization, your talk may be quite mathematical in nature (areas such as control theory and information theory come to mind). Or you may simply have a few equations localized on a couple of slides or sprinkled among your bullet lists, graphs, charts, diagrams, and photographs. Either way, we think it is sensible to break the issue of mathematical content down into main subissues: the equations on your slides, and what you plan to say about them.

The Equations Themselves

Appropriate Use of Notation

Let's start with the obvious: an *equation* must contain an *equals sign*. The following are *expressions*, not equations:

$$V^2/R, \qquad \frac{me^4}{2\hbar^3 n^2}\left[1 - \left(\frac{n}{n+1}\right)^2\right], \qquad \int_0^\infty f(t)e^{-st}\,dt.$$

The one on the far right, for example, could be used to write an equation defining the Laplace transform of a function $f(t)$:

$$F(s) = \int_0^\infty f(t)e^{-st}\,dt.$$

But all by itself it is *not* an equation. One should never be overtly sloppy with mathematical symbols or terminology.

Example. Here are some examples of mathematical carelessness from student slides:

$$M_s = 4\pi M_s \qquad\qquad w = \sum_{n=}^N x_n$$

$$f = \sum_{n=1}^N [n^2 + 2n^3 \qquad\qquad \int \cos x = \sin x$$

$$F \to ma \qquad\qquad f(\theta) = \cos(theta)$$

Because a talk is less formal than a formal document, however, some additional latitude does exist. In a formal written document, we would insist that the equals sign be kept out of non-formula text. Even on a slide, we would never condone

Concrete is strong and $=$ the best option

But, as clarity and efficiency are the main requirements for text on slides, we would assent to a sensible and consistent use such as

$t =$ time variable
$s =$ Laplace transform variable
$f =$ given signal
$F =$ Laplace transform of f

While slides are not necessarily about formality, they are still about effective communication.

How Many Equations?

Our answer is "not too many, but possibly more than you will actually refer to in the talk." The density of equations on your slides depends on a number of factors.

Example. Anita is talking to the local chamber of commerce about health effects from the fields produced by cell phone towers. Although she wishes to discuss topics like Specific Absorption Rate, which could easily be explained to engineers using a few equations, she knows her audience is not well-versed in mathematics and decides not to include any equations at all.

Example. Jose recently graduated with his Ph.D. in mechanical engineering and is preparing a short presentation that he will give to a group of staff scientists and engineers during a job interview. His goal is to convince the audience of his technical prowess rather than to educate them on the subject of his graduate thesis, so he chooses to include far more equations than he can cover, and plans to highlight the main ideas that they support. Being skimpy with technical detail could cause the selection committee to question the depth of Jose's knowledge on his thesis topic.

Most often some choice in between these extremes is best. At a technical conference, for instance, you might include several equations that are immediately recognizable by the audience, and require little description, but are needed as a lead-in to the novel contributions of your work or to define various terms.

What You Plan to Say about Your Equations

One of the hallmarks of seasoned technical speakers is their handling of equations during a presentation. Here are a few pointers to avoid coming across as an amateur. Considering them now could help you in the initial build stage of your presentation.

1. Don't read your equations symbol by symbol. The audience can see what the equation *says*; you should be telling them what it *means*.

Example. While displaying an equation such as

$$U_{\text{ave}} = \frac{1}{4\pi} \iint_{\Omega} U(\theta, \phi) \, d\Omega$$

you could simply say:

The average radiation intensity is given by this expression.

You might even add

Recall that the integrand function is the radiation intensity along a specific direction.

But don't point to each symbol and say:

U sub ave equals one over four pi times the double integral over omega of U of theta and phi d omega

without an especially good reason.

2. Don't believe you must refer to every equation you display. Your equations shouldn't control your presentation — *you* should. There are various reasons for glossing over an equation during a presentation. Maybe your audience is intimately familiar with the equation and no explanation is needed. Or maybe you are running short on time. Or perhaps you included the equation as insurance against confusion, but recognize that extra clarification is not needed.

3. Try to encapsulate the essential meaning of an equation. Is the equation challenging to solve, or is it relatively routine? Does it show that a quantity of importance is large or small? Does it show that something is increasing or decreasing? Does it show that a process is optimal, or that a system is stable? An equation can take a large effort for a listener to digest. Why should he or she care about this particular one?

4. Be logical. If certain equations lead to other ones, and if that's important, then say so. Don't confuse the audience by skipping around so that the connection between mathematical ideas becomes obscure.

5. Don't forget the English part. Certain words must be used to explain mathematics properly in English: *substitute, simplify, solve, recast, manipulate, invert, differentiate, integrate,* and so on. Terms such as these are seldom synonymous: *simplify* does not mean the same thing as *substitute,* for instance.

> **Example.** Consider substituting the value $x = 3$ into the equation $f(x) = x \cos(x)$. Don't say:
>
> I solved $f(x)$ for $x = 3$.
>
> Instead, say:
>
> I evaluated the function f at $x = 3$.
>
> The words *solve* and *evaluate* have different meanings.

4.6   Visual Aids

Here we consider the things you'll bring along to help you communicate with the audience. These may include slides, videos, equipment demos, etc.

Slides

As visual aids, slides can serve a number of purposes:

1. Provide a means to organize the talk. It might be difficult for a speaker who is not well trained or experienced to work without organizational aids.

2. Provide transitional aids and visual cues about content, for both the speaker and the audience. You don't want to forget anything important, so write it down!

3. Efficiently display specific technical information — graphs, charts, diagrams — for the purpose of information transfer to the audience.

4. Provide the audience something to engage with. This is helpful for speakers who have difficulty establishing rapport with an audience.

5. Provide an archive of the material presented. Often the slides constitute the only record of the talk (be it a briefing, pitch, interview talk, or a conference presentation that does not have an associated proceedings document). This is a good reason to put care into your choice of material for inclusion on your slides.

What Should My Slides Look Like?

The overall feel of a presentation is largely determined by the look of the slides. This is where you can lend a personal touch to your presentation, but you may be bound by certain restrictions. For instance, your company or the conference organizers may specify a required template. If you have complete freedom, you may want to start out using one of the many sample templates provided by commercial presentation software packages, or find something you like on the web and duplicate it. We list below several considerations for what to include.

We suggest you follow the old adage and *keep it simple*, but also *keep it clean and uncluttered*. An example slide created by the authors in Microsoft PowerPoint® is shown in Figure 4.4, and took just a few minutes to make. We use black, white, and gray for our examples, but you will certainly want to use a color background with perhaps a watermark, gradient, or some pleasing pattern (more about color below).

How Many Slides?

There is no magic rule regarding the number of slides to prepare. You are carrying out a design process and must consider what would be best for your purposes. Remember that your main goal is to communicate a certain amount of information to your target audience.

Even a presentation without slides is possible.[3] Slides are *visual aids*; this implies they are used — if at all — to help the speaker connect with the listener. You needn't make slides for the sake of slides, or 20 slides for the sake of 20 slides. You especially don't want to find yourself racing through slides just to "get through" everything you brought along. Nor do you want the slides to take over the presentation (or provide an ineffective "crutch" by giving you the opportunity to read them verbatim). So prepare what seems like a reasonable number of slides for the main part of your talk. The rehearsal phase (Section 5.1) will give you a chance to adjust your slides (both in number and in content).

---

[3]Remember, people have been giving engaging talks for millennia without using a single slide. Story-tellers have passed along oral history, philosophers have described difficult subject matter, and religious figures have explained detailed history and concepts, all without slides.

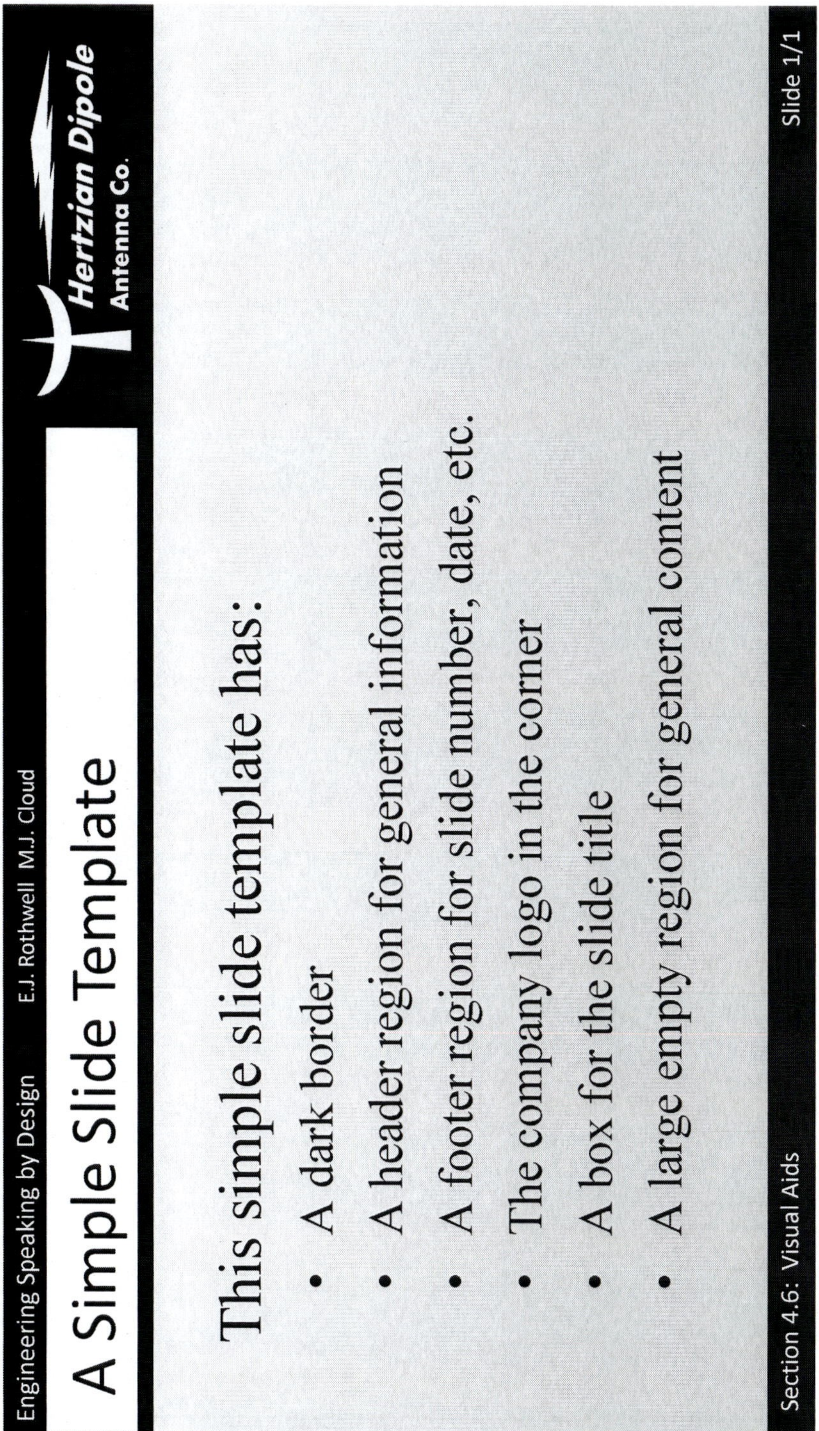
Engineering Speaking by Design
E.J. Rothwell M.J. Cloud
A Simple Slide Template
Hertzian Dipole
Antenna Co.
This simple slide template has:
• A dark border
• A header region for general information
• A footer region for slide number, date, etc.
• The company logo in the corner
• A box for the slide title
• A large empty region for general content
Section 4.6: Visual Aids
Slide 1/1

FIGURE 4.4
A simple PowerPoint® slide template created for this book.

Also keep in mind that it may be wise to design some auxiliary slides for use during the question-and-answer part of your talk. You may never use these, but it's better to be safe than sorry.

Example. Busy managers encourage their engineers to keep the number of slides in their presentations small, both to minimize the time spent in scheduled briefings and to lessen the burden of reviewing the slides. Recall that presentation slides often represent the only physical record of a briefing or idea pitch, and should be prepared carefully with knowledge of their archival value. Many organizations are stressing the use of just a few information-intense slides, with many visuals (graphs, pictures, flowcharts) and few words. Some groups employ a single briefing slide, usually broken into four panels, often called a *quad chart*.

Using just a few sparsely worded slides puts the burden on the speaker to know his material well. JoAnne adopted a strategy to help with her monthly briefings. She first created a presentation with many words and visual cues. These helped organize her thoughts and make sure she omitted nothing important. After rehearsing the presentation several times to herself, she cut the number of slides by half, keeping most of the visual information but eliminating slides that contained mostly words. She rehearsed again, eventually paring the number of slides to four. She checked that all essential information was still present on the slides so that someone reviewing them later could retrieve the underlying points and any important supporting data. JoAnne's boss and her colleagues appreciated her new presentation style. Her briefings seemed more spontaneous; no longer fixated on following the word structure of her slides, she was better at engaging her audience.

The Content of a Slide

As a component of your designed presentation, a particular slide should have a definite purpose. It may contain text, equations, a table, a chart, a graph, a diagram, a photograph, or a combination of these elements. There are few universally applicable rules about the content of a slide. A classic one is not to crowd any given slide with too much material. It is safe to say that an audience should never be required to read anything over a few words in length. Long figure captions, long paragraphs, etc., should be avoided.

Example. Alexander's biggest worry about his upcoming presentation was that he might forget to say something important. So he filled his slides with everything that might be needed. His plan was to formulate his points during rehearsal and adjust on the fly according to audience reactions. Unfortunately, slides crammed with material led to overconfidence and Alexander didn't prepare as well as he should have. When the big day came he resorted to reading directly from the slides. The listeners felt overwhelmed by the long paragraphs of tiny text, and Alexander's monotonous recitation quickly caused them to disengage.

Alexander's slide might have looked like Figure 4.5. The audience is left to wonder whether they should be reading everything they see, and the speaker is tempted to recite verbatim. A better alternative is shown in Figure 4.6. Only the main points are included, which Alexander can use as cues and the audience can use as a helpful guide. Of course this requires significant rehearsal, so that each point may be elaborated on as appropriate.

There is a difference between a slide with a lot of content and one with "too much text." Recall JoAnne's experience showing that a lengthy and detailed presentation can be made from a single slide with dense content. Space on presentation slides is a form of real estate worthy of much consideration, and there are few hard and fast rules. Use this as a guide, however: *never include the trivial or the obvious.*

Example. Jeremy displayed a picture of a three-terminal electrical device during his presentation. The very simple diagram was labeled "three-terminal electrical device." The listeners, all electrical engineers, felt insulted. Yes, they knew he was speaking about electrical devices. And yes, they knew how to count to three!

Slide Layout

This is where either a natural eye or an awareness of some graphic design fundamentals can be helpful. The latter subject covers issues of visual unity, balance, contrast, proportion, color, etc., and we recommend a few books in Further Reading on page 135. The pointers below are merely the results of our informal observations of slides made by engineers.

1. Again, don't crowd a slide with too much material. Graphic designers talk about *white space* and why it's important to have some. Do whatever you can keep your slide clean and clutter-free. A slide should contain a workable amount of information. If the presentation must be done on one slide with four panels, you will have to work extra hard to keep it readable.

2. Choose a legible font. The world of electronic fonts is constantly changing, so it is impossible to give a fixed recommendation about font choice. But you should certainly avoid exotic or cutesy fonts. And be sure your fonts are large enough to be deciphered from the back row of seats.

3. Make careful decisions about the use of color. Color can certainly add much interest to your slides. But in a formal technical presentation, color

Engineering Speaking by Design   E.J. Rothwell   M.J. Cloud

Hertzian Dipole
Antenna Co.

Example: Too Much Text!

These are the <u>attractive properties</u> of leaky-wave antennas

- Leaky wave antennas are much cheaper than phased array antennas. Phased array antennas require separate phase shifters and attenuators which must often be integrated into the antenna aperture – this is expensive!

- Leaky-wave antenna may be constructed on simple planar surfaces like circuit boards. This avoids needing cumbersome 3-dimensional structures which can cause a lot of problems.

- Leaky-wave antennas may be bent or shaped to be conformal to the surface of many types of vehicles, such as aircraft or ground vehicles. This helps improve the aerodynamic performance of the vehicles, and reduces the potential for damaging the antenna. The electrical properties of the antenna might be changed, though.

- Leak-wave antennas can be made with a higher gain than lots of types of antennas such as patches. The gain depends on the length of the leaky-wave antenna in wavelengths.

- Leaky-wave antennas are broadband due to the propagating wave nature of the current on the antenna. They may also be scanned in the elevation plane by changing the frequency within the broad band.

- Leaky-wave antennas are easy to fabricate using simple circuit-board techniques. They may be etched right onto the surface of the board and connectors may be soldered directly to the board.

Section 4.6: Visual Aids   Slide 1/1

FIGURE 4.5
Too much text for one slide!

Engineering Speaking by Design   E.J. Rothwell M.J. Cloud

Hertzian Dipole Antenna Co.

Slide 1/1

Example: Nice Amount of Text

Attractive Properties of leaky-wave antennas

- Low cost alternative to phased arrays
- Planar
- Conformal
- High gain
- Broadband (frequency scanning)
- Easy to fabricate

Section 4.6: Visual Aids

FIGURE 4.6

A clean slide containing the main ideas as cues for the speaker.

should be kept under control. Don't run wild with color for its own sake. Ideally, color should be there to communicate something of value (red for *hot* or *warning*, for instance). And remember that certain color combinations are garish, unappealing, or just plain difficult to see (such as yellow text on a white background). Also, keep in mind that a non-negligible fraction of your audience may be fully or partially color blind. This is a good reason to use color as an information *enhancer* rather than an information *carrier*.

4. Try to group related elements. If an equation and a figure are related, perhaps they could appear on the same slide (as long as the slide is not crowded as a result).

5. Understand that a slide may look different on a projection screen than it does on your computer screen. All your layout decisions should be regarded as provisional, subject to change after the rehearsal phase.

6. Consider the use of templates. Many commercial software packages, such as Microsoft PowerPoint®, have standard templates to help you lay out the various elements of your slides. Many other templates are available on the web. Your organization may have specific templates, either required or suggested. Conferences often have a required format or uniform template.

7. Consider the use of graphical tags. Small graphical elements may be added to each slide to identify your organization, the conference you are speaking at, or the project you are working on. These are usually placed at the corners, tops, or bottoms of the slides.

8. Include informative titles. Titles quickly convey to the audience the content of the slide. Use a large font and consider separating the title from the rest of the slide with a line, box, or other graphical element.

9. Don't forget the possibility of footers. The footer space at the bottom of the slide can be used for a variety of purposes. Consider including the slide number (for audience reference), the date, your name, the name of the speaking venue, or other significant information. This is useful if individual slides are extracted later from an archive of your presentation.

10. Be cautious with backgrounds. Presentation software generally allows you to select a background image, pattern, fixed color, or color gradient. These may enhance the look of the presentation if used subtly. Avoid garish backgrounds or backgrounds that hide or distract from your slide material. When using an image or watermark, be sure it is faded sufficiently to be discreet.

11. Don't go wild with animation and sound. Subtle animation, such as a text box opening up, can help to engage the audience. However, objects that fluctuate, spin, dance, vibrate, explode, or do many of the other visual tricks provided by presentation packages can be annoying. Remember that you are giving a professional presentation. Similarly, avoid cartoon-like sound effects. An exception is a movie or other animated element used to convey information.

12. Be careful about actively linking to web pages. Web pages are notoriously transitory. If you plan to click on a web link during your presentation, be sure the page still exists before starting the talk.

> **Example.** Tammy was a member of the internet generation. Her earliest reading experiences were associated with colorful web pages, pop-up advertisements, and flashy videos. So, she naturally assumed a good technical presentation must be equally splashy, with much color, animation, different fonts, varying text position and size, and lots of graphical elements. What she eventually learned is *everything in moderation*. A little flash and an occasional splash of color catches the audience's eye, but too much is just plain distracting. It is the speaker's job to keep the audience's attention through an engaging speaking style and by emphasizing the interesting nature of the material.

If you were in the audience and saw the slide in Figure 4.7, how would you react? In contrast, consider how Shreya's engineering eye quickly figured out a better approach.

> **Example.** As an undergraduate researcher, Shreya had watched many graduate students give presentations to her research group. Now that she had graduated and been accepted as a masters student, she was expected to similarly brief the group on her work. She had taken many notes during the past year on how students presented, the content of their slides, and the reactions of the group members — particularly her faculty adviser. She saw that the group was most receptive when the speaker concentrated on just one or two main points, including background information and recent accomplishments. She constructed several clean slides, each with just a few bullet points, an illustrative graphic showing the machine she was studying, and a plot or table summarizing recent results. She provided additional relevant information orally, emphasizing the important points with energy and enthusiasm. Her careful notetaking paid off; she was praised by her group mates and her adviser for a coherent and informative report.

Clear Equations

Among the most technical elements that appear on slides, and the hardest for audience members to digest, are mathematical equations. The utmost care should therefore be taken to produce logical and visual clarity. Positive examples of equation layout can be seen in practically any engineering textbook (but not necessarily on web pages). In addition to such obvious issues as font size, pay close attention to how multi-line equation displays are treated. Notice, first, that an effort is made to align equality signs vertically.

FIGURE 4.7
This slide is a mess!

Example. This is a nice multi-line equation display:

$$y(t) = \int_{-\infty}^{\infty} h(\lambda)\, x(t-\lambda)\, d\lambda$$

$$= \int_{-\infty}^{\infty} A\, e^{-\lambda/\tau}\, U(\lambda)\, B\, U(t-\lambda)\, d\lambda$$

$$= AB \int_{0}^{t} e^{-\lambda/\tau}\, d\lambda$$

$$= AB\,\tau(1 - e^{-t/\tau}).$$

Note the vertical alignment of the equality signs.

Then, if an expression must be broken because of excess length, the break occurs at a sensible spot.

Example.

$$P_e = \frac{1}{2} \int_{-\infty}^{V_T} \frac{1}{\sqrt{2\pi}\sigma_0} \exp\left[\frac{-(r_0 - s_{01})^2}{2\sigma_0^2}\right] dr_0$$

$$+ \frac{1}{2} \int_{V_T}^{\infty} \frac{1}{\sqrt{2\pi}\sigma_0} \exp\left[\frac{-(r_0 - s_{02})^2}{2\sigma_0^2}\right] dr_0$$

$$= \frac{1}{2} \int_{(-V_T + s_{01})/\sigma_0}^{\infty} \frac{1}{\sqrt{2\pi}} \exp\left(\frac{-\lambda^2}{2}\right) d\lambda$$

$$+ \frac{1}{2} \int_{(V_T - s_{02})/\sigma_0}^{\infty} \frac{1}{\sqrt{2\pi}} \exp\left(\frac{-\lambda^2}{2}\right) d\lambda$$

$$= \frac{1}{2} Q\left(\frac{-V_T + s_{01}}{\sigma_0}\right) + \frac{1}{2} Q\left(\frac{V_T - s_{02}}{\sigma_0}\right).$$

Sensitivity with respect to details like these is what leads to an overall work-manlike effect. Your equations will be easier to present and easier for the audience to grasp.

As with text, there is a temptation to read equations verbatim. If the speaker is not a diligent rehearser, he may resort to saying

"... we use the curl of E equal to minus d B d t equation in the next equation of the integral from a to b of E dot d L ..."

Instead, the speaker should provide context for the equation:

"... substituting Faraday's law into the expression for voltage allows us to ..."

If the speaker emphasizes the meaning and importance of the equations, it is possible to include quite a few on a slide — as long as the audience is not overwhelmed by the sheer density of mathematical symbols. Figure 4.8 shows a slide used by one of the authors at a technical conference. While the slide seems to contain a large number of equations, the author emphasized their relevance to the problem being analyzed. He did not dwell on the particular mathematical form of each equation. He also assumed the audience had a basic familiarity with the symbols and an understanding of their meaning, and included a graphical element to identify geometrical parameters. Mathematical rigor is always important; there is value in mathematical detail when the presentation is archived for later reference.

Readable Diagrams and Graphs

Engineers often convey information through diagrams, graphs, and tables. The key difference between using these items in a presentation versus a written document is the time the user has available to glean information. A reader is able to dwell over a complex diagram for as long as it takes to extract the needed information. During a presentation, he may have only one minute to do the same. This may be impossible, even under the guidance of the speaker.

Key on these two things: *readability* and *complexity*. An audience member must be able to see and understand the graphic element from the far end of the room (just as she must be able to hear your words). *Readability* means that the shapes must be clear, the fonts large, and the lines bold. *Complexity* implies that you should not overwhelm the listener with content. How will he react to a diagram with a hundred elements? Probably by tuning out.

Both diagrams and graphs must follow the same rules of readability and complexity, so let's concentrate on graphs. A *bad* graph might have any of these characteristics:

1. Fonts too small (readability). Can your text be seen from the back of the room?

2. Color used poorly (readability). Be careful when combining colors. Contrast is essential. Do not, for instance, put a yellow line against a white background.

3. Lines too thin or not differentiable (readability). Thin lines are hard to see. Data lines must be distinguishable in some way. Use different line types, but never so many that you cannot tell them apart.

4. Background too busy (readability). Data lines are difficult to see when placed against a patterned background.

5. Too many graphs on a page (readability and complexity). You may want to put several graphs on a page so that you can contrast and

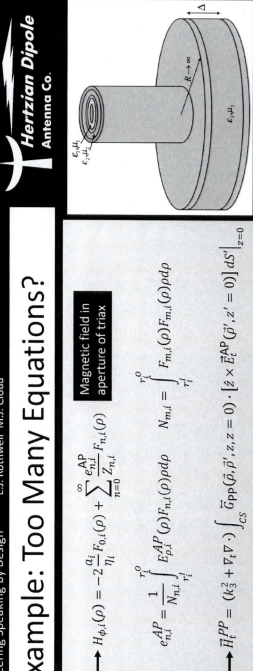

FIGURE 4.8

A slide with a lot of mathematical detail. Don't read the equations verbatim!

compare results. This can be done, but only with great care. Small graphs must be balanced by fonts that appear unusually large and by lines that appear unusually thick. Unless kept particularly simple, multiple graphs on a slide can quickly overwhelm the audience.

6. Captions too long (complexity). Don't include too much information in a caption. The audience may feel compelled to try to read the entire thing, which will distract from your message. Consider placing essential supporting information in a separate text box or in a table, either within or outside the area occupied by the graph.

7. Legends complicated or ambiguous (complexity). You may opt for a legend if you have several data lines on a graph. This can distract an audience member, who will have to correlate the line types, symbol types, or colors with the legend. If there are several lines, it may be hard to discriminate between them. Instead, include annotations next to the lines. This gives immediate contextual meaning.

Figure 4.9 shows a graph with many of these issues. Imagine viewing this slide from the back of the room! Contrast this with Figure 4.10. Considerations of readability and complexity provide visual benefits that are quickly apparent.

Should I Include Tables?

It is rarely beneficial to tabulate long lists of data. This type of data is best presented using a graph. However, smaller tables are fine as long as readability and complexity are adequately addressed. Compare Tables I and II in Figure 4.11. The data in Table I is overwhelming and hard to assimilate. This information could easily be plotted in a graph and thereby made accessible. In contrast, the data shown in Table II does not lend itself to graphical representation and is best tabulated. It could be quickly explained and contrasted without trying the listener's patience.

Demo Equipment

Although slide presentations seem to be the norm for engineers, there are times when only a live demonstration will do. Here are some pointers we've gained from experience.

1. Make sure the entire audience can see the demo. This may be easily accomplished if you are demonstrating the function of a new car window. But how are you going to demo a hand-held blood pressure monitor in an auditorium?

If the audience is large, consider using one or more cameras and a monitor or projection screen. The camera can be focused on a device, output display, input panel, or some visual aid. If you lack access to a camera, consider using a prerecorded video of the demo to show the larger audience while presenting

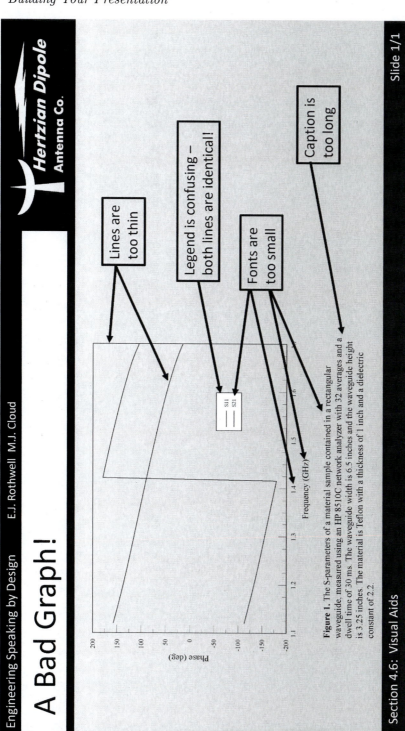

FIGURE 4.9
A bad graph.

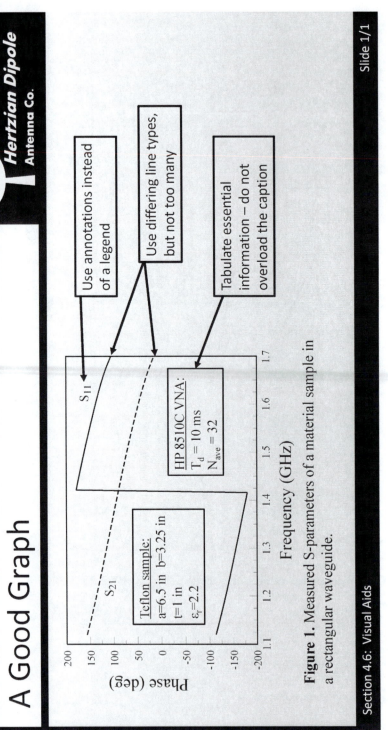

FIGURE 4.10
A good graph.

Engineering Speaking by Design   E.J. Rothwell M.J. Cloud

Using Tables in Slides

Hertzian Dipole Antenna Co.

Table I. Measured reflection coefficients for different material thicknesses

a/lambda	125 mil Gamma (real)	Gamma (im)	250 mil Gamma (real)	Gamma (im)
-1.66778	-0.277412612	-0.121712349	-0.744740035	-0.18285533
-1.501	-0.269928569	-0.117193274	-0.7251157	-0.175981199
-1.33422	-0.263179192	-0.11320888	-0.707399826	-0.169910146
-1.16744	-0.257057729	-0.109664825	-0.691314206	-0.164500563
-1.00067	-0.251477132	-0.106488151	-0.676632943	-0.159643187
-0.83389	-0.246365809	-0.103621531	-0.663170747	-0.15525229
-0.66711	-0.241664391	-0.101019268	-0.650774201	-0.151259589
-0.50033	-0.237323252	-0.098644461	-0.639315161	-0.147609961
-0.33356	-0.233300597	-0.09646696	-0.628685723	-0.144258363
-0.16678	-0.229560975	-0.094461871	-0.618794343	-0.141167588
0	-0.226074114	-0.092608439	-0.60956282	-0.13806597
0.166778	-0.222814009	-0.090889214	-0.600923925	-0.135649265
0.333556	-0.219758183	-0.0892894	-0.5928195	-0.133173425
0.500334	-0.216887109	-0.087796366	-0.585198978	-0.130860126
0.667111	-0.214183733	-0.086399256	-0.57801815	-0.128693061
0.833889	-0.2116309	-0.085088677	-0.571238188	-0.126658108
1.000667	-0.209221991	-0.083856464	-0.564824834	-0.124742967
1.167445	-0.206938764	-0.082695476	-0.558747738	-0.122936872
1.334223	-0.204773036	-0.08159944	-0.552979906	-0.12123035
1.501001	-0.202715557	-0.080562818	-0.547497237	-0.119615029
1.667779	-0.20075805	-0.079580703	-0.542278147	-0.11808348

Table II. Measured dielectric strength for various materials.

Material	Number of samples	Dielectric strength (kV/mm)
Fused silica	27	530
Polypropylene	33	24.1
PVC	25	13.9
Neoprene	16	20.1
Porcelain	63	127

Section 4.6: Visual Aids   Slide 1/1

FIGURE 4.11
Using tables in slides.

the demo live to a smaller audience (of judges, for example). Cameras can also be used to record the demo or stream it to the web.

2. Audience safety is paramount. The authors have witnessed capstone design presentations involving projectiles, whirling saw blades, noxious gases, sparking electrodes, laser emissions, and so forth. Your listeners should not have to worry about their personal safety while attending your talk.

3. If possible, have a backup unit with you. Remember Murphy's law and its various corollaries. Bring extra batteries if the unit is battery powered.

Example. Marshall was thrilled when the device he invented won the student design competition sponsored by his professional society. The best part was that he got to attend a European conference where he would demonstrate his prize-winning contraption. Trying to set his project up at the conference center, Marshall found that the wall outlets differed from those in the USA. He wasn't too worried, since his demonstration wasn't until later that day, and asked around until he found someone with an adapter plug. He connected everything to his satisfaction, flipped the switch, and *pow* — his power supply belched smoke and died a horrible death. Only then did he learn that Europe uses 220 volts, while he had designed his power supply for 110 volts. Unable to fix his contraption, Marshall was forced to talk about how it would work if it had electricity. A little homework would have saved much embarrassment.

Posters

Poster presentations are now common at conferences and review meetings. The speaker is usually expected to stand beside a large visual aid (a poster) and interact one-on-one with interested viewers. Here are some things to keep in mind while preparing your poster and giving your presentation.

1. Be aware of formatting requirements. The allowable poster size is determined by the table or easel on which the poster is mounted. The authors have seen beautifully prepared posters drooping ingloriously because they were too large for their supports. There may also be specific requirements about the title size, the fonts, the organization of the material, or the number of panels. Some conferences dissuade presenters from using multiple sheets of paper. Others even permit multiple easels.

2. Control the amount of content. Authors may be tempted to compress an entire oral presentation onto a poster. This generally leads to a cluttered, visually unappealing, confusing mess. Omit any details you can provide to an interested viewer during discussion, and concentrate on main ideas.

3. Spatial layout is important. Posters differ from oral presentations because all aspects of the presentation are available to the viewer at once. This can be overwhelming unless some spatial structure is present. One approach is to replace the temporal sequence of slides in an oral presentation by a spatial sequence of ideas. Use clearly delineated panels with titles to guide viewers through the sequence. Another approach is to organize the poster around carefully placed graphical elements. This is appropriate if the order of information access is not important.

4. Be conscious of visual aspects. Many of the same concerns regarding slides are appropriate here. Format graphs and equations properly. Avoid lengthy captions and large tables. Be careful with color and font selection. Be sure that everything is easily readable by someone standing several feet from the poster (don't annoy the viewer by requiring her to stoop over and squint at microscopically small fonts).

5. Know the rules. You may be required to install your poster at a certain time and remove it before the next poster session. There may be a specified time period during which you are required to attend your poster to answer viewer questions. Be sure to wear the attire expected of oral presenters.

6. Consider ancillary elements. It might be helpful to bring electronic copies of your poster to distribute to interested viewers. You might also wish to distribute a business card or photograph of your product, system, or experiment. Consider bringing a demo unit or other physical items if allowed.

4.7 Ethics

To what extent must a speaker be concerned with ethics? Isn't it true that, unlike the written word, the spoken word has only temporary effects? Don't you believe it!

Example. Jim took liberties and made a false statement to get a questioner off his back. Jim wasn't aware that others in the audience were recording his talk on their smartphones.

In fact, speaking is just as serious as writing. All of professional communication is serious; practically by definition, professionals have access to specialist knowledge that the general public does not have. This places significant responsibility on you as an engineer. Any false statement can terribly mislead and do real damage (to yourself and others). One of the best things a presenter can do is to continuously monitor whether he or she is speaking in an ethical fashion. Consider this dimension while building your materials; it's a necessary part of vetting them for readiness.

Like logic, grammar, mathematics, and graphic design, the subject of ethics can easily fill a whole book (we list a few on page 138). Fortunately, the various engineering professional societies and organizations have done a reasonably good job of distilling classical wisdom into their official *codes of ethics*. Although an ability to recite the code published by your own organization would be ideal, even a passing familiarity with the ethical code published by any of the major engineering societies would provide you with some valuable insurance as a technical speaker. The codes published by the Institute of Electrical and Electronics Engineers (IEEE) and the American Society of Mechanical Engineers (ASME) represent excellent starting points for consideration. The IEEE Code of Ethics, for example, contains provisions against conflict of interest, bribery, falsifying data, and taking credit for the work of others. The very act of joining the IEEE means that you agree to these provisions.

It should go without saying that there are frequent and significant intersections between ethical codes and the laws of various countries. People who have been granted security clearances must be extremely cautious with their written and spoken communications. But engineering ethics takes on many forms, depending on the type of work involved. In the US research community, engineers are bound to act responsibly as described in, for example, the National Science Foundation's policy of *Responsible Conduct of Research*.[4]

Example. Shannon is a Ph.D. student watching her adviser give a presentation at a small local conference. He is describing recent experiments that Shannon conducted in their research laboratory. She notices that he is only describing the experiments that produced results which validate their working theory, and that several experiments that returned negative results are being overlooked. After the conference she confronts him and tells him of her discomfort over his "cherry picking" of results. He tells her not to worry — these are preliminary results and a better experiment will return improved data before they publish in the open literature. "Besides," he says, "this is a small conference and there is nobody really important here. And no one will remember what we showed anyway."

4.8   Checklist: Building Your Presentation

☐ **I have established the target specifications for my talk**
☐   I have established the scope of my talk
☐   I know what topics I want to cover
☐   I have thoroughly investigated the motivations and interests of the
    audience

---

[4]http://www.nsf.gov/bfa/dias/policy/rcr.jsp. Last viewed January 24, 2015.

☐ I understand the time limitations
☐ I understand the venue limitations
☐ **I have prepared my visual aids**
☐ I have selected appropriate presentation software
☐ *I know whether specific software must be used*
☐ *I am familiar with the software or willing to learn*
☐ I have created a slide template
☐ *I know whether a format is required*
☐ *My template includes helpful information*
☐ Title
☐ Page number
☐ Graphical tag (e.g., organizational logo)
☐ Footer or header with date/venue/presenter etc.
☐ *My template is clean and uncluttered*
☐ I have decided on the number of slides to use, based on time limit and audience needs
☐ I have checked the appearance of all my slides
☐ *Fonts are appropriate*
☐ I did not mix too many fonts
☐ The size is sufficient to see from all points in the venue
☐ *The amount of material on each slide is appropriate*
☐ The number of equations, tables, and figures on each slide is appropriate
☐ I did not combine too many elements
☐ *I did not use too much text*
☐ There is not more text than the audience can easily read
☐ I only include enough text to provide myself cues and give the audience a point of focus
☐ *Color is appropriate*
☐ I chose color carefully
☐ I did not mix too many colors
☐ Color combinations are distinctive
☐ I considered the impact of color on audience members who are color blind
☐ *Slide background is appropriate*
☐ The background is not too dark
☐ The background is not too busy
☐ The background is not distracting
☐ *Adjunct elements are appropriate*
☐ Animation is not overbearing or otherwise distracting
☐ Sound or movies add to the impact of the presentation and are not just for show

☐   Active links to web pages do not disrupt the flow of the presentation

☐   *My equations, graphs, and tables are easy to read*

☐   Font size is not too small

☐   Lines are not too thin

☐   Captions are short, easy to read, and meaningful

☐   Legends are clear and easy to understand

☐   There are not too many elements on any slide

☐   *I have checked the appearance of my presentation using the venue equipment or similar equipment*

☐   The presentation is visually appealing

☐   All fonts, equations, and graphs appear as expected

☐   All adjunct elements (animation, links to web pages, sound, movies) behave as expected

☐ **I have carefully examined my materials for any logical blunders**

☐   There are no fallacies of formal logic

☐   There are no fallacies of informal logic

☐ **I have checked my materials for errors in technical claims**

☐   My claims are reasonable

☐   My claims pass dimensional and order-of-magnitude checks

☐   My claims agree with known special cases

☐   My claims are in accord with known standard principles

☐   My evidence comes from reliable sources

☐   I have examined relevant counterexamples to my claims

☐ **I have checked my materials for proper English usage**

☐   My word choice is appropriate

☐   *Words have the proper technical meaning*

☐   *Words are at the proper level of formality*

☐   The punctuation and spelling obey standard rules

☐   My wording takes the form of either description or argumentation, as appropriate

☐   *My descriptive content has a clear standpoint*

☐   *My descriptive content has the appropriate level of detail (no more, no less)*

☐   *My descriptive content forms a coherent picture*

☐ **I have checked my materials for proper mathematical discourse and clarity of presentation**

☐   I have used proper mathematical notation

☐   I have not overburdened the listener with too many equations

☐   My equations form a logical progression

☐   I understand what I want to say about each equation

☐ **I have carefully planned my demo**

☐   The demo is easily seen by all audience members

- ☐ There are no safety issues
- ☐ I have a backup plan in case of loss or damage of the demo equipment
- ☐ **I have considered the special needs of a poster presentation**
- ☐ I am aware of the formatting requirements
- ☐ I know the rules of presentation
- ☐ I have determined the amount of material I can place on the poster
- ☐ I have created a workable layout
- ☐ *I have replaced the temporal orientation of an oral presentation with an appropriate spatial orientation*
- ☐ *I have created an alternative layout that is clear and easily understood by the viewer*
- ☐ I have created or gathered needed ancillary items (handouts, business cards, brochures, etc.)
- ☐ **I have considered any ethical implications of my presentation**
- ☐ I understand what is appropriate ethical conduct within my discipline/profession
- ☐ I have consulted the ethical codes for my organization
- ☐ I understand the special needs of my audience

4.9 Chapter Recap

1. The target specifications for designing your talk must consider the nature and scope of the subject material, the nature and size of the audience, the time alloted, and the characteristics of the venue.

2. Logic, grammar, mathematics, graphic design, and ethics are all key aspects of quality control when it comes to a technical presentation.

3. Learning about fallacious arguments is one effective way to avoid making them.

4. Simple set diagrams can be helpful in spotting formal logical fallacies.

5. It only takes one counterexample to invalidate a formal syllogism.

6. Know all the terms on your slides (what they mean and how to pronounce them!).

7. Be consistent with English punctuation.

8. Don't confound the listener with a muddle. Deliver well-planned descriptions and logically sound arguments.

9. Time spent learning English grammar is valuable even for speakers who do not plan to implement every rule they encounter.

10. Mathematics is about clarity and precision. It is rarely the place for abuse of symbols or terminology.

11. Don't plan to read your equations symbol by symbol. In fact, you can show certain equations on your slides without referring to them at all. For the most part, however, you should provide engineering-level meaning to your equations.

12. Design your slides carefully. They should be clear and inviting visual aids. There is no definite rule regarding the number of slides you should prepare.

13. Special considerations are required for a safe and successful equipment demo.

14. Be aware of the formatting and presentation requirements specific to poster presentations.

15. Engineers are professionals and are subject to ethical requirements.

4.10 Exercises

4.1. Identify the argument as valid or fallacious.

(a) Gold has a density of 19,300 kg/m³. This material has a density of 19,300 kg/m³. Therefore, this material is gold.

(b) All elements with an atomic number of 79 are gold. This element has an atomic number of 79. Therefore, this element is gold.

(c) Some metals are malleable. Gold is malleable. Hence gold is a metal.

(d) Nobel prizes are made from gold. Some engineers have Nobel prizes. Therefore, some engineers have gold.

4.2. Assume P, Q, R, S are statements. Are the following argument forms valid? Explain.

(a) P or Q.
         If P, then R.
         If Q, then R.
         Therefore, R.

(b) P or Q.
         If P, then R.

If Q, then S.

Therefore, R or S.

(c)   Not-R or not-S.

If P, then R.

If P, then S.

Therefore, not-P.

(d)   Not-R or not-S.

If P, then R.

If Q, then S.

Therefore, not-P or not-Q.

4.3. Consider the following list of *invalid* categorical syllogisms. Find a real-world counterexample that reveals each syllogism as invalid.

All P is M.	All M is P.	All M is P.
Some S is M.	No M is S.	All M is S.
$\therefore$ Some S is P.	$\therefore$ No S is P.	$\therefore$ All S is P.
No M is P.	Some P is M.	Some M is not P.
All M is S.	Some S is M.	Some S is M.
$\therefore$ No S is P.	$\therefore$ Some S is P.	$\therefore$ Some S is P.
Some M is P.	No P is M.	Some M is not P.
Some S is not M.	No S is M.	No S is M.
$\therefore$ Some S is not P.	$\therefore$ No S is P.	$\therefore$ Some S is not P.
No M is P.	No M is P.	All M is P.
All S is M.	Some S is M.	All S is M.
$\therefore$ All S is P.	$\therefore$ Some S is P.	$\therefore$ No S is P.
All M is P.	Some M is P.	All M is P.
Some S is M.	All S is M.	Some S is M.
$\therefore$ Some S is not P.	$\therefore$ All S is P.	$\therefore$ All S is P.
No M is P.	Some M is not P.	Some M is not P.
Some S is M.	All S is M.	All M is S.
$\therefore$ No S is P.	$\therefore$ No S is P.	$\therefore$ No S is P.

For example,

All P is M.	All cats have four legs.
Some S is M.	Some chairs have four legs.
$\therefore$ Some S is P.	Therefore some chairs are cats.

4.4. Look for fallacies.

(a) Jones's results are questionable because she made significant errors in the past.

(b) Since it is impossible to conceive of anything but overheating causing this problem, the problem must be due to overheating.

(c) Having discovered failures in two of the connectors tested at random, we concluded that all 5000 connectors likely failed.

(d) The vibration level increased after we heard the noise, hence the noise must have caused the vibration increase.

(e) Transistor leads are like tiny legs. Since insects have six legs, transistors have six leads.

(f) Only two possibilities exist: either the flux decreased or it increased. Since both of these represent changes, we do know that the flux changed over time.

(g) This motor is superior to the other alternatives because it is better.

(h) Resistors often have white stripes. Therefore, they seldom have green stripes.

(i) All machines are somewhat inefficient. Zach is somewhat inefficient. Hence Zach is a machine.

(j) A capacitor is an electrical device. A transistor is an electrical device. So a capacitor is a transistor.

(k) There must be something wrong with subsystem Q. Ever since it was installed, subsystem R has been unreliable.

4.5. As humans, we have *cognitive biases* that lead us to distort our experiences and process information selectively. Do some background reading about cognitive biases. Could any of these patterns make it easier to commit fallacies?

4.6. Consult a logic textbook (e.g., one of the books listed on page 137) to learn the Venn diagram method for validating syllogisms. Use the method to validate the following 19 syllogisms:

All M is P.	No M is P.	All M is P.
All S is M.	All S is M.	Some S is M.
$\therefore$ All S is P.	$\therefore$ No S is P.	$\therefore$ Some S is P.
No M is P.	No P is M.	All P is M.
Some S is M.	All S is M.	No S is M.
$\therefore$ Some S is not P.	$\therefore$ No S is P.	$\therefore$ No S is P.
No P is M.	All P is M.	All P is M.
Some S is M.	Some S is not M.	No M is S.
$\therefore$ Some S is not P.	$\therefore$ Some S is not P.	$\therefore$ No S is P.
Some M is P.	All M is P.	Some P is M.
All M is S.	Some M is S.	All M is S.
$\therefore$ Some S is P.	$\therefore$ Some S is P.	$\therefore$ Some S is P.
Some M is not P.	No M is P.	No P is M.
All M is S.	Some M is S.	Some M is S.
$\therefore$ Some S is not P.	$\therefore$ Some S is not P.	$\therefore$ Some S is not P.
All M is P.	No M is P.	No P is M.
All M is S.	All M is S.	All M is S.
$\therefore$ Some S is P.	$\therefore$ Some S is not P.	$\therefore$ Some S is not P.

All P is M.
All M is S.
$\therefore$ Some S is P.

Note that care is required with the last four of these in cases where empty sets are permitted. The syllogisms in the sixth row only hold if the class M has at least one element. The last entry only holds if P has at least one element.

4.7. Check the following equations for dimensional correctness.

(a) The kinematic relation $x = x_0 + v_0 t + \frac{1}{2}at^2$, where x and x_0 are distances, v_0 is a speed, a is an acceleration, and t is time.

(b) The kinematic relation $v^2 = v_0^2 + 2a(x - x_0)$, where the quantities are as in part (b).

(c) The kinematic relation $v = (2gh)^{1/2}$, where v is a speed, g is the free-fall acceleration constant, and h is a height.

(d) The energy balance relation $\frac{1}{2}kx^2 = \frac{1}{2}mv^2$, where k is a spring constant, x is a displacement, m is a mass, and v is a speed.

(e) The rotational kinematic relation $\theta = \omega_0 t + \frac{1}{2}\alpha t^2$, where θ is an angle in radians, ω_0 is an angular speed, α is an angular acceleration, and t is time.

(f) Bernoulli's equation $v^2/2 + gz + p/\rho = $ constant, where v is a speed, g is the free-fall acceleration constant, z is a height, p is a pressure, and ρ is a density.

4.8. Some groups of words are easily confused because, for instance, they sound alike. For example

- their, there, they're

- lead, led

- we're, where, were

- sight, site, cite

- all together, altogether

- its, it's

- passed, past

Make a list of words that you personally find confusing.

4.9. Engineering speakers are not the only ones embarrassed by improper word choices or by usage blunders. The Internet is rife with hilarious examples appearing in newspaper headlines and on signs. Make a list of your favorites. Take particular note of any you think you might have made yourself (if you hadn't done this assignment first!).

4.10. Classify the following passage as description or argumentation.

Let P and S designate the primary and secondary planes, respectively, of the transformer. Our goal is to reflect the secondary circuit to the primary side In this case we may replace the circuit to the right of plane P by its Thevenin equivalent. Consider the open-circuit terminal voltage. Since $I_1 = NI_2$ and $I_1 = 0$, we have $I_2 = 0$. This implies that the voltage across the secondary plane is $V_S = V_{s2}$, hence $V_{oc} = V_{s2}/N$ across the primary plane.

4.11. Repeat for the following passage:

First let us assume thermal equilibrium with no externally applied voltage. Charge-carrier density gradients are set up by diffusion processes that occur across the junction region when it is formed. The immediate junction area consists of a depletion region characterized by a lack of mobile charge carriers; an associated electric field constitutes a built-in potential barrier with a contact potential of a few tenths of a volt (commonly 0.7 V). No net current flows across the junction in equilibrium; there is a precise balance between diffusion currents and "drift currents" caused by the junction field. The depletion region is also known as the space-charge region. It is around 0.5 μm thick, and has a junction capacitance typically in the low-pF range.

4.12. In addition to advancing sound arguments and providing clear descriptions, engineers have to develop and explain *classifications*. For example, chemical engineers classify aqueous solutions as *acid* or *base*, according to pH, but may also choose to further classify acids as either *strong* or *weak*, or as either *organic* or *inorganic*. Generate a slide that serves to classify some objects or ideas of interest to you. Practice the presentation to yourself, anticipating possible questions or objections.

4.13. On page 76 the following were cited as examples of mathematical carelessness. Precisely what is wrong with each one?

(a) $M_s = 4\pi M_s$

(b) $w = \sum_{n=}^{N} x_n$

(c) $f = \sum_{n=1}^{N} [n^2 + 2n^3$

(d) $\int \cos x = \sin x$

(e) $F \to ma$

(f) $f(\theta) = \cos(theta)$

4.14. Write out your favorite equation from physics or engineering, and practice explaining it without reading it symbol by symbol.

4.15. Linda is a sales engineer for ABC Corporation. Her job requires technical knowledge, but she is really the person that sells the product to her clients. One of Linda's clients asks her to speak at his son's high school. Are any ethical considerations involved here? Why or why not?

4.16. Build a presentation and construct a rubric to evaluate the result. You may wish to use the checklist from Section 4.8 as a guide. Attend the presentation of a speaker that you admire, and apply the rubric.

5

Optimizing Your Presentation

Having built your presentation and its associated materials, it's time to optimize prior to the big event. The basic approach is to rehearse, evaluate, tweak your materials, and repeat.

5.1 Rehearsal

By making one or more test runs through your talk, you will have the chance to gauge your fluency with presenting the subject, your ability to fit the time limit, your chances at connecting with the audience, your ability to use various aspects of the venue to your advantage (or at least cope with them), and even your readiness for some likely questions from the audience.

Readiness to Present the Subject

One goal of rehearsal is to make sure you're ready to present the subject of your talk. We break this down into two main issues.

1. You will need an adequate grasp of the subject matter. Let's start with the *terminology* of the subject: how well do you know it? If you are planning to use a word, do you know what it means well enough to provide a definition if asked? Are you able to support that definition with an example or reinforce it with an illustration if necessary?

> **Example.** Gloria is giving a presentation about her design methodology to a local group of professional engineers. She describes how she uses a MATLAB® optimization toolbox to find "the optimal solution to our problem." She uses the word "optimal" several times until someone raises her hand and asks "how do you know the solution is optimal?" "Well," says Gloria, "it meets our requirements, and anyways it would take too long to find a better solution." The audience member goes on to explain that the word "optimal" has a specific understood meaning, and the meaning is

not "good enough." A bit flustered, Gloria realizes she could have avoided this criticism with a proper choice of words.

Next, what about the *big ideas*? A sure way to alienate listeners is through a barrage of details with no technical context. They will want to have (indeed, will often *need* to have) some context into which the details can fit.

Example. Baozhi is working to improve the brake pads for a high-performance automobile. He needs to brief the design group on his proposal to change the way in which the ceramics in the pads are sintered. He knows that before discussing the details of the material processing, he must put the process in context. "I know you are all used to the prime failure mechanism being thermal overload," he begins, "but the problem we're dealing with here is failure due to vibration." Given context for the following technical discussion, the audience is prepared to understand why the steps in the process do not correlate with their expectations of structural failure.

This does not imply, of course, that the details are not important. They are, and they must be presented well. After a good technical presentation, the audience will come away with knowledge of essential pieces *and* their interconnections.

2. You will need an ability to deliver the subject matter. This is *not* the same as merely having an adequate grasp of the subject in your own mind, as many novice presenters have learned the hard way.

Example. Recall our presenter Brett on page 69. Brett certainly knew that an *algorithm* is simply a set of steps for solving a problem. But when the time came to actually say the word *algorithm*, he struggled and finally gave up. That's not an experience he will strive to repeat!

Mere pronunciation is not enough, of course, which is why we reviewed some points on logic and English description in the last chapter. You must be ready to deliver a clear description of your system, device, algorithm, or idea. You must also be ready to reason from premises to conclusions, or from causes to effects.

Readiness to Fit the Time Limitation

While staying on schedule is conceptually simple, going overtime seems to plague the experienced speaker almost as often as the novice. Start by making sure you know what the time limit is. Don't *assume* you can meet it; do a test run using an actual clock or timer (if your presentation is part of a moderated program, the moderator will certainly have one handy). And be aware that *it is easy to misjudge time while delivering a technical presentation.*

Example. A moderator will often give a subtle warning, such as a wave of the hand, when the speaker has nearly exhausted his time. If the speaker finds that he is not on schedule, the tendency may be to panic and jump to the end. This is what Warren did when the moderator flashed the "3 minutes" sign during his talk. He said, "Gosh, I guess we're almost out of time. I'd better skip to the end." He read through his conclusions while the audience sat wondering what they had missed. If this happens to you, try to make a quick assessment of what remains and give the audience a brief summary of what you'll be skipping. Finish with an encouraging suggestion to speak with you after the session for more information.

What if you discover that you are going long? During your talk, you *may* be able to fix this by increasing your tempo, by paying attention to, and eliminating, any *filler* expressions ("um, whatever, uhhh, OK, er"), or by reducing your intended level of explanation. (Understand, however, that you should never speak faster than the practical "information absorption rate" of the audience.) It's much better to discover an overlength issue during rehearsal. You then have the option to redo your outline or redesign your slides with increased time efficiency in mind. For instance, you could adjust the breadth or depth of coverage to save a minute or two. Although you might not be happy tweaking a "perfectly designed" presentation, it's better than being interrupted mid-sentence by a moderator with "I'm sorry Mr. Jones, your time has expired."

Readiness to Connect with Listeners

An ability to connect with audience members depends on many things, including eye contact, body language, movement, responsiveness, and enthusiasm.

Example. Most people understand the importance of making a connection with the audience, but many don't realize how that connection can be broken by odd speaking habits. Sean spoke regularly at a series of noontime technical seminars at his company, and engineers appreciated having access to his expertise on many topics. However, after one seminar a colleague took him aside and told him that his habit of saying, "Well, how about that!" and rubbing his hands together every few minutes had become a local joke, and attendees were starting to take bets on how many times he would do it during one of his talks. In fact, the habit was so distracting that the engineers weren't paying attention to what he was saying but instead were on the edge of their seats waiting to tally the next occurrence of his odd habit. Sean made a point to become aware of every time he made his trademark statement, and managed to reduce the number of incidences to once or twice a talk. The audience, perhaps at first a bit disappointed, soon became more attuned to the content of his presentations.

Readiness to Perform in the Venue

Is your speaking volume adequate for the size of the room? Will you have to use a microphone? If so, do you know how to use one? Is your computer system (both hardware and software) compatible with the projection system at the venue? When does the presentation need to be loaded onto the computer? Will you be using a pointer (such as a laser pointer) and have you practiced holding it steady? Is there a formal dress requirement and do you have everything needed? (It's not a bad idea to actually rehearse in the clothes you will be wearing during the real presentation, if they are clothes you are not accustomed to wearing.) Will your talk be recorded? There are *many* issues you could consider, depending on the sort of venue you are facing.

Readiness for Questions

How should you prepare to field questions?

1. You need an extensive knowledge of your topic. There is obviously no substitute for this. Few things are worse than a technical presenter who has a marginal grasp of his or her subject.

2. You need an awareness of the boundaries of your knowledge. No one will be able to know *everything there is to know* about a topic, and it is likely that you will be asked a question to which you do not know the answer. It's impossible to be ready for every conceivable question, just as it's impossible to conceive every question in advance. However, it is possible to know your limits. In fact, it is *essential* to know your limits.

3. You need a willingness to be forthright about your limitations. Beyond knowing your limits is a willingness to *acknowledge* your limits. Nothing exposes the underprepared, self-conscious speaker quite like an attempt to bluff his way through a tough question. Don't regard saying "I don't know" as an admission of ignorance, but rather an indication of openness to learning something new. Perhaps the person asking the question can help educate you on the topic.

4. Prepare a list of outside resources to suggest in case you are uncertain how to answer anticipated questions. There will be areas where you know that your knowledge is weak. Prepare a list of experts in the field who could help, or a list of good books, technical articles, or dependable websites to suggest.

5. Consider creating a personal algorithm for handling questions. We outline a basic one in the next chapter.

Example. Molly scheduled a job interview with her dream employer, where she was required to give a short seminar describing her master's degree work. Her talk went well, but afterwards an engineer raised his hand to ask a question.

> All of your work assumes a linear model. Most of our products operate in the nonlinear region. What happens to your system in the nonlinear region?

Molly paused for a moment and responded.

> My work is meant to be used with a very different system than your company uses and it was not necessary to consider nonlinearities. I honestly don't know what will happen in the nonlinear region, but there have been a lot of studies on related nonlinear systems, mostly done by Johnson and Keers, I believe. We could start from there if you're interested in applying my work. Let's talk later.

Impressed by Molly's poise, honesty, and self-confidence, the engineering manager hired Molly and had her adapt her techniques to the systems used by the company.

Readiness to Keep a Firm Hold on Things

There may be times when you sense the audience attempting to hijack your presentation. Yes, the audience is the essential element of every presentation, but you — the speaker — are in charge. It's up to you to keep a firm hold on the situation, especially when subject to a stringent time limit.

Here are a few things novice presenters do to lose control.

1. They get interrupted by nonstop questions, or spend too long interacting with a single audience member. In a 15-minute presentation you may simply not have time to take questions during the main part of the talk. In a longer talk you may, but you should avoid any temptation to turn your talk into a public conversation between you and a dominant, aggressive, or particularly persistent audience member. Find a polite way to say, "Let's talk alone, later."

2. They let the murmuring go on for too long. You want the audience to be affected by your talk. At times, however, a brief spurt of laughter or gasp of surprise can morph into dozens of private conversations. This means you got their attention with something, which is good. But you can't afford to lose their attention — along with precious time and your own mental flow — by waiting too long for them to calm down. So don't be afraid to re-assert yourself. Never let your presentation degenerate into a general social hour!

3. They try to distribute handouts during the talk. Handouts can be made available before or after the talk, or both, or via a web address. By expecting the audience to pass around a stack of papers during the presentation, you create distraction. People get occupied worrying about when it might be their turn to get their handout, or who they are supposed to pass the stack to next, and they stop paying attention to you. Don't give the audience an excuse to disengage from your presentation.

> **Example.** Tanya has been asked by her boss to give a briefing on her redesign of a headlight assembly. She begins by saying, "The new airflow design allows the bulb to operate ten degrees cooler than before, extending its life by 100 hours. I've been asked to describe the retrofit for the body mounts." She is quickly interrupted by Andrew, who works on the wiring harness for the headlight assembly. He's been bothered for some time about the design of the harness connector. "That's all well and good, but I want to talk about the wiring harness." Everyone in the audience begins to roll their eyes, because Andrew brings this up at every meeting. "We're not here to talk about the wiring harness," responds Tanya. "A lot of people have come today expecting to hear about the headlight assembly." Andrew is persistent, and Tanya finally ends the distraction by offering to arrange a personal meeting for him with Luke, the project leader.

5.2   Feedback and Evaluation

Feedback regarding the quality of your performance level could be derived from several sources. The first is your own sensory awareness. Try to monitor your speaking volume, clarity, rate, and so on. This takes practice and can be difficult at first, but the alternative is to neglect significant information about how you are doing. The second is an assistant if you're lucky enough to have one. A colleague who has experience speaking and also some familiarity with your technical area would be ideal. But even a friend or family member can provide valuable feedback on the nontechnical aspects of a practice session. Finally, you could record yourself for later playback and analysis.

Example. Roger gave his 15-minute talk to his cousin Mick for practice. Roger felt ready when everything seemed to go smoothly. He was certainly perplexed to hear Mick say, "Everything was fine except that you keep slapping your thigh. Why are you doing that?" *Impossible! Slapping my thigh? What on earth is he talking about?"* But Mick wasn't dreaming and had the video recording to prove it. Every time Roger was done pointing at the screen, he would drop his hand and let it hit his right thigh with a slap. Roger decided to practice more in order to rid himself of this distracting habit before his real presentation the following week.

5.3 Revision and Iterative Improvement

As a result of rehearsal and subsequent evaluation, you may be realizing a need to change certain things. We have already examined some typical and important possibilities:

1. *subject grasp*: base knowledge, detailed familiarity, access under pressure

2. *materials*: slides, pointer, apparel, demo equipment

3. *delivery*: rate, volume, eye contact

4. *venue*: acoustics, lighting, projection equipment

Now is the time to make changes. After that, rehearse again and repeat the iterative improvement cycle until no additional changes are deemed necessary.

5.4 Checklist: Optimizing Your Presentation

☐ **I have rehearsed thoroughly**
    ☐ I have rehearsed in front of an audience member or colleague
    ☐ I have received feedback from knowledgeable listeners
    ☐ I have adjusted my presentation based on the feedback
☐ **I completely understand the material I will be presenting**
    ☐ I understand the terminology
    ☐ I understand the technical nature of my material
    ☐ I understand what others in my field have done
    ☐ I understand how my work fits with what others have done

☐   I understand how my work fits within my discipline
☐ **I am mentally ready to deliver a clear presentation**
☐   I know how to pronounce the terms
☐   I am comfortable with the material
☐   I have identified material that requires special care to explain
☐ **I have my timing down**
☐   I know the time limit
☐   I have identified portions of the presentation I stumble over, and have taken extra care with these passages during rehearsal
☐   I have prepared for possible interruptions during my talk
☐   I have an "exit strategy" if I take longer than I prepared for
☐ **I am prepared for the venue**
☐   I have a pointing device
☐   I have adjusted the volume of my speech to accommodate the room
☐   I have the appropriate clothing
☐   My talk is in the proper format for the presentation equipment
☐   My demo is set up properly for the room
☐ **I am prepared to answer questions**
☐   I am knowledgeable of, and comfortable with, my material
☐   I understand the limits of my knowledge and when to say "I don't know"
☐   I have prepared a list of resources that will be helpful to audience members
☐ **I am mentally prepared to control the audience**
☐   I understand how to deal with unruly or insistent audience members
☐   I am willing to limit my interactions with any given audience member
☐   I understand how to "calm" a distracted audience

5.5   Chapter Recap

1. Rehearsal is a highly recommended practice for public speaking preparation.

2. A good technical presenter has the depth of knowledge required to clarify technical terminology with concrete examples and vivid illustrations.

3. The audience will need the big picture as well as the details in order to grasp your subject.

4. It is easy to lose track of time while giving a technical presentation.

5. Many factors go into connecting with listeners, but primary among these is eye contact.

6. An important aspect of preparation is readiness for a specific venue.

7. It is important to be ready for questions from the audience. A speaker may even wish to *over-prepare* on the technically difficult aspects of a topic. Tough questions can come from people who are interested in the topic as well as those who (for whatever reason) want to make you look bad.

8. While the speaker has the floor, he or she is supposed to maintain control of the situation.

9. A friend or a video camera can provide valuable perspective on your rehearsal talks.

10. Iterative improvement is a key aspect of preparation for a speaking engagement.

5.6 Exercises

5.1. Name a few attitudes that you deem desirable in a technical presenter.

5.2. Name a few attitudes that you might expect to encounter in audience members.

5.3. List your greatest fears about public speaking.

5.4. Can you think of a way to estimate the number of words you will say in a 20-minute presentation?

5.5. Compile a list of words that you have difficulty pronouncing. Practice saying these words in a variety of situations.

5.6. Discuss how one might deal with *preconceived notions* in an audience.

5.7. It was Leon's habit to insert the words "to be honest" before certain facts:

To be honest, we haven't looked into that yet.

Comment on this habit. Name some other distracting habits you have observed in technical presenters.

5.8. Describe your reaction as a listener when the speaker

(a) begins the presentation with an anguished sigh

(b) points out an error on one of his or her own slides (e.g., "Actually, that should be MHz, not kHz.")

(c) stops speaking suddenly and stares at his or her slides for half a minute without saying anything

(d) continually refers to the "right side," which is in fact the audience's left side

(e) uses humor of a personal nature (e.g., "Fluid velocity is also important when you take a shower ... which hopefully you all did today.")

(f) seems confused about the topic of his or her talk

(g) expresses doubt about his or her grasp of the topic (e.g., "I've been trying to figure out how this works exactly, but haven't had much luck.")

(h) acts as if he or she is being coerced into giving the presentation

(i) opens the presentation with a vague generality (e.g., "Magnetism has long been of interest to humankind.")

These behaviors have been observed by the authors many times during student presentations.

5.9. Construct a rubric to evaluate how well you have optimized your presentation. You may wish to use the checklist from Section 5.4 as a guide.

5.10. Attend the presentation of a speaker that you admire, and apply the rubric developed in Exercise 5.9.

6

Showtime: Delivering Your Presentation

You've spent many hours preparing, tweaking, and rehearsing your presentation. Now the big day has arrived; all of your hard work is about to pay off. But remember, as for a great athlete or musician, preparation is only half the story. You still have to perform. Whether this is your first talk in front of an audience or your two hundredth, you'll need to give it your all to achieve the lasting impact that is the goal of all meaningful presentations.

6.1   The Countdown

Before leaving for the venue, check and double-check your materials. If possible, get there early.

> **Example.** Tom had done his homework regarding the classroom in which he was scheduled to speak, visiting it a week beforehand. Fortunately, however, Tom arrived a bit early on his presentation day and discovered sunlight streaming through a big window right into his eyes. After some fiddling with clunky window shades, the situation was much improved. Tom had the situation under control before anyone else arrived.

Now *relax*. You are as prepared as the whole situation, to this point, permitted you to be. So when the time comes, just take a deep breath and ...

6.2   Do It!

Things to remember:

- Keep breathing.

- Face the audience.

- Eye contact.

- Appropriate volume.

- Rate of speech.

- Enthusiasm, some controlled movement, gestures.

- Communicate (don't read your slides).

- You have the floor!

What makes a person a strong presenter is that their presence shines through, showing their expertise and passion for their topic.

Example. Sandy had several speech classes and remembers being told to "maintain eye contact!" However, it wasn't until she was giving a talk to a local high school group that she realized how much more this phrase means than just looking at people. While scanning the audience, Sandy noticed a girl who perked up and smiled each time she made eye contact. She decided to look to this girl whenever making a key point, to judge the impact of her message. Afterwards the girl came up to Sandy and said "I can't tell you how much I enjoyed your talk — it was like you were speaking directly to me! It's tough being a girl who likes engineering. I was thinking of trying something else in college, but your talk was so inspiring that I think I will stay with it! Thanks!" Sandy then realized that the true purpose of eye contact is to make a real human connection with the audience.

6.3   If You're Nervous

Try these tips:

1. Envision yourself as a successful presenter. Harness the psychological power of positive visualization. The image you have of yourself in your mind's eye can have a huge impact on your performance.

2. View your audience as friends and colleagues. This can make you instantly more comfortable. You don't have to regard your listeners as menacing strangers! At a minimum, they are probably engineers or at least persons interested in some aspects of engineering activity.

3. Don't tell the audience how nervous you are. They don't need to know that. Chances are good that you look much better than you feel on stage anyway.

4. Focus on your message and information. This way you'll place less of your focus on yourself.

5. Forgive yourself for glitches. We'll have more to say about this important practice in Chapter 7.

6.4 When the Unexpected Occurs

In a word, *adapt*. Roll with it! Keep smiling. Remember your primary job and do your best to get the information across regardless of what just happened. And don't say anything you'll regret afterwards.

> **Example.** Aldo had recently come up with a change to a manufacturing process that would save his company quite a bit of money. He was presenting his concept to the division head and several managers when his boss, Henry, started to get agitated and glance repeatedly at the division head. Eventually Henry stood up and said "I told you that we have a constraint that must be met and you're not considering it!" Aldo knew that Henry had never mentioned the constraint and was probably trying to save face with the division head. Rather than confront Henry and say something that he would regret, he immediately altered the direction of his presentation and said "OK, we can deal with that. Let me show you how this would work under that constraint." It took some finesse and quick thinking, but Aldo was able to outline a rough plan. The result impressed everyone in the room — including Henry.

If nothing else, you'll have a fun story to tell later on; everyone understands how challenging it is to speak in front of an audience, and they enjoy the occasional story illustrating Murphy's laws.

> **Example.** Matthias regularly presented his work at a regional conference and thought he knew the drill. Presenters were expected to bring electronic copies of their slides and place them on the computer in the auditorium at the start of their session. Being somewhat overcautious, Matthias always brought copies of slides on two different USB memory sticks. He was a bit perturbed this day when the computer refused to read the first USB stick. "Well, good thing I have a backup," he thought, but the computer refused to read the second stick as well. As the first speaker in line, Matthias needed a quick solution to his predicament. Scanning the room, he noticed an old-fashioned overhead projector in the corner. He was able to salvage the situation by writing his presentation on the projector in real time,

copying from a printed practice set of his slides. The talk was well received; furthermore, Matthias now had a great example of "if things can go wrong, they will."

6.5 Handling Questions

Here is a simple algorithm for accepting questions from an audience. First, recognize the audience member if they have their hand up.

Example. Yes, a question?

Now listen carefully to the question and decide whether you understand it fully. If not, ask for clarification before attempting to answer.

Example.

Audience: What about kids playing nearby?

Speaker: OK, are you asking whether the system will be safe for children?

If you do understand the question, then repeat it back to the audience before saying anything else. It is often hard for people sitting in the back row to hear another listener pose a question.

Example.

Audience: What kind of approximation did you use there?

Speaker: The question pertains to the kind of approximation I used here.

At this point, you must decide whether and how to answer the question. It is possible that the question would take too long to answer,. You certainly have no obligation to answer a question that is more appropriate for a lengthy discussion after your talk is finished.

Example.

Audience: Can you outline for us the proof of Hubert's theorem?

Speaker: The listener has asked me to outline the proof of Hubert's theorem. I'd love to do that, but unfortunately there's not enough time. However, I'd be happy to discuss a proof after the session, or provide a reference for you via email.

The question could be too far outside the scope of your presentation.

Example.

Audience: Can you tell us the curing time of the polymer you used?

Speaker: The question pertained to the curing time for our polymer. Well, as you know, curing time depends on many variables that are determined by the process employed. Since our choice of process hasn't been made yet, and isn't really relevant to this discussion, the specification of curing time is outside the scope of my presentation. I'd be happy to talk with you after the session about some processes that our group is considering.

The question may not be answerable (by you, at least) at this time.

Example.

Audience: Can existence of a solution to equation (10) be established?

Speaker: The question is whether existence of solution can be proved. Unfortunately I don't know the answer to that at this time. But it is something that interests us and we're currently working on it.

It is possible that only a partial answer can be given within the time constraints.

Example.

Audience: Can you outline for us the proof of Hubert's theorem?

> Speaker: The question is whether I can outline how to prove Hubert's theorem. A full outline would take us too far afield, but the basic idea is that we subdivide our region of interest and apply equation (5) in each tiny subdivision. It turns out that all the interior contributions cancel, leaving us with the theorem statement.

Finally, it is possible that a full answer can be given. If so, it is good policy to ask whether the answer was understood and found satisfactory by the listener.

Example.

> Audience: What kind of approximation was used to obtain equation (12)?

> Speaker: The question is in regards to the method of approximation used to get equation (12). The binomial approximation was used there. That's why the exponent conveniently disappeared in the next line. Is that a sufficient answer to your question?

6.6   Checklist: Delivering Your Presentation

☐ **I have double-checked that I have all my necessary materials**
☐ **I have given myself sufficient time to travel to the venue**
☐ **I have put myself in the proper frame of mind**
  ☐ I have gotten sufficient sleep
  ☐ I have eaten a good meal
  ☐ I have allowed some time before the talk to 'de-stress' and become calm
☐ **I am giving myself feedback during the talk**
  ☐ I am maintaining eye contact
  ☐ I am speaking with enthusiasm and using appropriate gestures
  ☐ I am speaking with appropriate volume
  ☐ I am aware of my pace and the time remaining
  ☐ I am aware of my personal quirks
    ☐ *appearance*
    ☐ *stance and movement of arms or hands*
    ☐ *overly repeated words or phrases (you know, like, hey, moving forward)*
    ☐ *pause fillers (um, uhhhh, throat clearing)*
☐ **I am prepared to adapt**

☐   I have an exit strategy
☐   I have rehearsed with an audience
☐ **I have a strategy for handling questions**
☐   I repeat questions after they are asked
☐   I ask for clarification if needed
☐   I defer questions that would take too long to answer
☐   I admit when I don't know the answer
☐   I am prepared to offer suggestions of outside references

6.7   Chapter Recap

1. Any speaker, regardless of their venue or purpose, has to engage the attention of the audience.

2. Early arrival at the venue can give you a chance to address unforeseen problems.

3. Relax, keep breathing, eye contact ... you have the floor!

4. Don't stand in front of your visual aids; let the audience see.

5. Talk to the audience, not to your visual aid.

6. Adapt, roll with the unexpected, keep smiling.

7. Have a plan for handling questions.

6.8   Exercises

6.1. Outline strategies for handling the following Q&A situations.

(a) Someone asks a question in a hostile tone of voice.

(b) Someone crosses professional boundaries and asks you a personal question.

(c) Someone asks a question that you, in effect, have already answered.

(d) No one asks a question.

6.2. Prepare a 15-minute technical presentation on a topic of your choice.

6.3. Construct a rubric to evaluate how well you have prepared to deliver your presentation developed in Exercise 6.2. You may wish to use the checklist from Section 6.6 as a guide.

6.4. Construct a rubric for evaluating how well you deliver your presentation from Exercise 6.2. You may wish to use the checklist from Section 6.6 as a guide. Deliver your presentation while a colleague uses the rubric to evaluate your talk. Discuss the results during a debriefing after your talk.

6.5. Make a list of your personal quirks based on the results of Exercise 6.4. Deliver your presentation again and see if your evaluation improves.

6.6. Attend the presentation of a speaker that you admire, and apply the rubric developed in Exercise 6.4.

7

Success! (What Now?)

It took a lot of hard work to get to this point. You had to design and prepare your presentation. You had to prepare yourself so that you could deliver the talk with confidence and poise. And now it's over. Is there anything left to do?

7.1 It's Over

Relax. Take a deep breath. Congratulate yourself for putting in your best effort. You took an intelligent, thoughtful approach to preparing yourself and your presentation, and hopefully everything paid off. If you did as well as you dreamed you might, and the accolades poured in, then relax, take it easy, and bask in your glory. You are in a special class of gifted speakers and can skip to the next section. If, on the other hand, the road was a bit bumpier and a thing or two didn't go as planned, consider these important points.

1. You did your best under the circumstances. You followed a sound preparation methodology, put all your engineering skill and experience to work implementing that methodology, thoughtfully *designed* a presentation for this particular audience (or at least the audience you expected to get), carefully built the presentation, optimized it through rehearsal, and then did your best to deliver it faithfully. That's something to be proud of.

2. You're human. Engineers are human, and humans make mistakes. That's the way it is. You don't have to be happy about it, but recognize that you're not alone. Also recognize that there is *always* room for improvement when human beings are involved.

3. The audience was human. Audience members sometimes say and do stupid things. They can be agendized and aggressive. It's quite possible that whatever incident has you so tied up in knots is not entirely your fault. Or maybe not your fault at all. Try not to be judgmental at this point. Don't mull it over, but instead sleep on it. It's likely that the listeners thought you did great and you are being way too hard on yourself. Ask one or two of them for feedback while they're still in the room; you might be surprised by the positive things they say.

4. You can't do any better than your best. Again, you did your best. *You can't do any better than your best at a given time.* Meditate on that fact if you're really feeling bad about how things went. Then resolve to go on improving your skills.

> **Example.** Camilla rehearsed her talk for the big convention about a million times. This was her first time speaking at such an important venue and she was determined to impress her colleagues. She had the timing down perfectly: 15 minutes of the allotted 20, leaving 5 minutes for Q&A at the end. Unfortunately, when the moderator gave the "2 minutes" sign, Camilla found that she had only completed half her talk. She scrambled to finish the rest and in the process consumed most of the Q&A time. She could have considered her presentation a flop. Instead she used it as a learning experience, adjusting her rehearsal methods in order to better predict how long she would need in front of a live audience.

7.2   Chatting after the Talk

People may approach you immediately after the presentation. Some may have questions they weren't comfortable asking in front of everyone else, or for which there was not sufficient time. Others may simply say thanks for what you did. Regardless of the forms these interactions take, they may represent chances to help others or to make new professional contacts. Try to stay open to the possibilities. If you're a working engineer, consider having some business cards handy.

> **Example.** Kyle was finishing graduate school. The pressure to find a job was mounting. After presenting his research findings at a conference, Kyle was thrilled that Jorge (a big name in the field) came up to chat. They talked about Kyle's work, which Jorge considered "ingenious." Kyle offered his business card, letting Jorge know that he was seeking a job. In addition to Kyle's contact information, the card displayed a web address for a site that he had created to describe his background and research. A few days later Kyle got an email from Jorge: "I don't hire MS students, but I know a lot of people who do. I was very impressed with your work and took the liberty to contact a couple of folks in industry. You may be hearing from them soon." Kyle's presentation was an opportunity to impress. He took full advantage of it, and of the subsequent opportunity to make important contacts.

Unless you are a student in a classroom setting where criticism is part of the experience, it is unlikely that an audience member will criticize your talk afterwards. It *could* happen, of course, but in our experience it probably won't. The chances are much greater that you will receive praise. But either way, honest feedback should be graciously accepted. Respond with a pleasant thanks for the information and file it away for later evaluation.

> **Example.** After his pleasant experience with Jorge, Kyle was approached by a second audience member. John criticized Kyle's techniques and suggested that a different method was better warranted. Kyle thanked John and asked for recommended references on the alternative approach. He also asked about John's willingness to speak again after he had a chance to examine the references. They exchanged email addresses; an initially confrontational interaction eventually turned into a fruitful collaboration of several months.

7.3 Use the Experience to Improve

Whether you came away from the talk exhilarated with the knowledge that you were on top of your game, or a bit discouraged from a sub-par performance, you can use your experience to improve. Take some time to assess how things went, and determine what you might have done to make things go better. Focus on three things.

1. Examine your performance. What did you do well and what flopped? Good speakers are constantly tweaking their speaking style.

2. Assess your preparation. Did you do everything necessary to prepare for the talk? What else could you have done? What would you do differently next time? Did you understand the audience as well as you thought?

3. Review the content of your presentation. Did you include everything you needed? Did you include too much, or the wrong material? What would you add? What would you remove?

If possible, discuss your presentation with an audience member. Their immediate reaction can be valuable.

> **Example.** Jordan was asked to brief a local group of civil engineers on a new technology for controlling traffic in construction zones. He rehearsed successfully in front of colleagues who were familiar with the technology.

Unfortunately, during the actual event, Jordan could tell from the questions that things didn't go as planned. His colleague Connie attended, and later they did a post-mortem over dinner. "There seemed to be a lot of confusion in the audience," said Jordan. "What do you think went wrong?" Connie was able to speak with several attendees and learned that their backgrounds in traffic engineering were not as broad as she and Jordan had anticipated. "They really are old school when it comes to maintaining traffic. They're especially unfamiliar with the newest reconfigurable systems. We should have had you review some earlier systems rather than jumping into the latest technology right away." Jordan realized that he had prepared well — but for the wrong audience. After this he changed his speaking style for similar events. If any confusion showed on the faces of audience members, he would stop and ask for feedback before continuing. He also brought some extra slides he could use if the audience seemed to require extra background information. Jordan understood that this approach might be impractical under tight time constraints. Even then, however, he found that the extra slides came in handy during one-on-one discussions after his talk was finished.

7.4   Prepare to Be in Demand

Skilled technical speakers are rare and, in many environments, highly valued. It is not unlikely that you will be asked to repeat a talk. This is a good thing: a request for a repeat performance means that someone saw value in the presentation.

Example. Kendall's boss Sophia was impressed by his monthly briefings to the platform teams. On Friday, the local congressman was coming to visit the plant for an update on the innovative techniques the company was using to reduce energy consumption. Sophia asked Kendall to give the presentation: "I know you've only been peripherally involved with green manufacturing, but I'm sure you can get up to speed on the main issues by Friday. You, among all our engineers, would have the best rapport with the congressman." Kendall was a bit apprehensive, but he prepared well and the update went off without a hitch.

If the request doesn't come from your boss at work, but rather from an interested audience member, then you needn't decide on the spot; you can graciously say thanks for the invitation, record contact information, and promise to be in touch about the idea later. You have every right to decline, of course.

But giving it some thought isn't a bad idea. You may be mentally, physically, or emotionally spent after your presentation and simply want to go home and forget the whole thing. You may, however, feel different about things in a few days. Much thought and effort went into the design, building, testing, and delivery of your presentation. Why not at least consider the possibility of giving it again? It may be a lot easier the second time. And there are always more people to help, more professional contacts to be made.

> **Example.** Oliver gave an impressive talk to his computer communications working group about how electromagnetic emission limits are set for wireless routers. He was approached afterwards by an engineer: "I work with a group educating the public on the safety of wireless systems, and next week we are meeting with some local folks who are worried about the recent proliferation of microcells in their community. Would you be willing to give your talk to them?" Aware that electromagnetic fields are a touchy issue with the public, Oliver asked for some time to think it over. Eventually he decided that giving the public an accurate account of this important health issue was worth the effort.

Of course, if you do decide in favor of a repeat performance, keep in mind that no two situations are exactly alike. If the venue is to be different the second time, you have some investigating to do.

7.5   Consider Reinventing Your Talk

Another possibility is to re-envision your presentation. A good presentation can be turned into a different but equally good presentation. The level could be raised or lowered; the scope could be broadened or narrowed. The subject could be shifted to a neighboring topic, with or without significant overlap.

> **Example.** Audience response to Oliver's talk on emission limits suggested a real need for addressing the public's concerns about electromagnetic fields. Oliver realized he could augment his talk with World Health Organization guidelines and National Institute of Health data, and produce a presentation with wide public appeal. Soon he was asked to speak to a number of civic groups and became respected for his honest assessment of radiation hazards.

7.6 Archiving Your Talk

If you did well, you must be proud of your presentation. Too bad so few people got a chance to view it! Why not get it out for the world to see?

There are several ways to archive a talk. If your organization has a web page, see if you can place a prominent announcement and a link to your slides (or even to video, if your presentation was recorded). If not, place it on your personal web page. Be mindful of material that is proprietary or otherwise restricted. If you belong to a professional society, ask whether there is an opportunity to post materials on their page.

Example. Gabrielle's talk to the local road commission about best practices for trail construction was well received. She asked if she could post her slides on their website and they agreed. Commissioners from neighboring counties read her slides, were intrigued by her ideas, and asked her to speak to them. Soon the word on the street ran to "If you're building a new trail, you've got to talk with Gabrielle first!"

You might also consider more permanent means of archiving your presentation materials. Why not turn them into a column for your company's newsletter or an article for a trade magazine? You could even write a technical paper for an academic journal. Better yet, why not use them in the first chapter of that book you've been wanting to write?

7.7 Checklist: After the Talk

- ☐ **I congratulate myself for all of my hard work**
- ☐ **I forgive myself for any missteps or blunders**
- ☐ **I engage with interested audience members**
- ☐ **I use the experience to improve**
  - ☐ I assess whether I did enough to prepare; what worked and what didn't?
  - ☐ I assess how well the talk went and how I can use the experience to improve my delivery
  - ☐ I assess whether the talk needs to be changed if I deliver it again
- ☐ **I consider reinventing my talk for another venue or occasion**
- ☐ **I find a way to archive my talk**

7.8 Chapter Recap

1. Not all presentations go well. It's always possible to learn from each mishap, though.

2. Informal chats with audience members can provide the speaker with new connections and opportunities for professional service.

3. Seek and accept constructive criticism as a way to improve. If you spend all your energy being defensive, you'll never grow.

4. Really good (knowledgeable, articulate, informative, entertaining) technical presenters are hard to find. The better you get, the more in demand you're likely to be.

5. The preparation for a good presentation need not go to waste afterwards. Many presentations can be repeated or adapted to new forms or occasions.

6. Consider archiving your talk, either on the web or in a permanent venue.

7.9 Exercises

7.1. Attend a presentation by someone you respect. Be a fly on the wall; watch how the speaker interacts with audience members after the talk. Make notes that will be helpful for the aftermath of your own presentations.

7.2. Make a list of the places on the web that you might consider archiving a presentation.

7.3. Construct a rubric to evaluate how well you handle yourself after your presentations. You may wish to use the checklist from Section 7.7 as a guide. Attend the presentation of a speaker that you admire, and apply the rubric.

Further Reading

Public Speaking

We mentioned that you can find many comprehensive books on public speaking. Some more recent titles include:

How to Write and Give a Speech: A Practical Guide for Anyone Who Has to Make Every Word Count, by Joan Detz. St. Martin's Press, 2014.

Public Speaking: An Audience-Centered Approach, by Steven A. Beebe and Susan J. Beebe. Pearson Education, 2000.

Public Speaking, 5th ed., by Michael Osborn and Suzanne Osborn. Houghton Mifflin, 2000.

The Art of Public Speaking by Stephen Lucas. McGraw-Hill, 2011.

Principles of Public Speaking, by Kathleen M. German, Bruce E. Gronbeck, Douglas Ehninger, and Alan H. Monroe. Longman, 2000.

Older but insightfully written books include:

Speech: A First Course, by E.C. Buehler and W.A. Linkugel. Harper & Row, 1962.

Speech: A Course in Fundamentals, by S. Judson Crandell, Gerald M. Phillips, and Joseph A. Wigley. Scott, Foresman and Company, 1963.

Titles written specifically for engineers and scientists include:

The Craft of Scientific Presentations: Critical Steps to Succeed and Critical Errors to Avoid, by Michael Alley. Springer, 2013.

Designing Science Presentations: A Visual Guide to Figures, Papers, Slides, Posters, and More, by Matt Carter. Elsevier, 2013.

Oral Communication Excellence for Engineers and Scientists: Based on Excecutive Input, by Judith Shaul Norback. Morgan & Claypool, 2013.

A broader book on general technical communication (including oral presentation) is

Essential Communication Strategies for Scientists, Engineers, and Technology Professionals, by Herbert L. Hirsch. IEEE Press, 2003.

Composing Mathematical Arguments

Books that focus on mathematical discourse include:

A Transition to Advanced Mathematics, by Douglas Smith, Maurice Eggen, and Richard St. Andre. Cengage Learning, 2010.

Introduction to Mathematical Structures and Proofs, by Larry J. Gerstein. Springer, 2012.

A First Course in Abstract Mathematics, by Ethan D. Bloch. Springer, 2011.

The Art of Proof: Basic Training for Deeper Mathematics, by Matthias Beck and Ross Geoghegan. Springer, 2010.

Reading, Writing, and Proving: A Closer Look at Mathematics, by Ulrich Daepp and Pamela Gorkin. Springer, 2011.

The Nuts and Bolts of Proofs, Fourth Edition: An Introduction to Mathematical Proofs, by Antonella Cupillari. Academic Press, 2012.

How to Read and Do Proofs: An Introduction to Mathematical Thought Process, by Daniel Solow. Wiley, 2009.

One way to hone reasoning skills is to study *sophisms*: arguments intended to deceive. See

Lapses in Mathematical Reasoning, by V.M. Bradis, V.L. Minkovskii, and A.K. Kharcheva. Dover Publications, 1999.

Mathematical Fallacies and Paradoxes, by Bryan H. Bunch. Dover Publications, 1997.

Engineering Design

For discussions of the engineering design process, we recommend:

Fundamental Concepts in Electrical and Computer Engineering with Practical Design Problems, by Reza Adhami, Peter M. Meenen, and Denis Hite. Universal Publishers, 2007.

Creative Engineering Design, by Brian S. Thompson. Okemos Press, 1996.

English Grammar, Style, and Vocabulary

The reader who is interested in English grammar and style could consult such books as

Engineering Writing by Design: Creating Formal Documents of Lasting Value, by Edward J. Rothwell and Michael J. Cloud. Taylor & Francis/CRC Press, 2014.

The Elements of Style, by William Strunk Jr. and E.B. White. Longman, 1999.

Errors in English and Ways to Correct Them, by Harry Shaw. Collins Reference, 1993.

The Most Common Mistakes in English Usage, by Thomas E. Berry. Mc-Graw Hill Professional, 1971.

Style: Toward Clarity and Grace, by Joseph M. Williams. University of Chicago Press, 1995.

Grammar for Journalists, by E.L. Callihan. Chilton Book Company, 1979.

We also recommend having an unabridged print dictionary such as

Random House Webster's Unabridged Dictionary. Random House Reference, 2005.

For some interesting historical background on English vocabulary, see

English Vocabulary Elements by Keith Denning, Brett Kessler, and William R. Leben. Oxford University Press, 2007.

Words Words Words, by C.M. Matthews. Macmillan, 1980.

Logic and Critical Thinking

For clear introductions to the principles of formal logic, we recommend

Introduction to Logic (14th Edition), by Irving M. Copi, Carl Cohen, and Kenneth McMahon. Pearson, 2010.

Logic for Mathematicians, by J. Barkley Rosser. Dover, 2008.

For less formal discussions under the heading of "critical thinking," see, e.g.,

How the Great Scientists Reasoned: The Scientific Method in Action, by Gary G. Tibbetts. Elsevier, 2013.

An Introduction to Critical Thinking and Creativity: Think More, Think Better, by Joe Y.F. Lau. Wiley, 2011.

Workbook for Arguments: A Complete Course in Critical Thinking, by David R. Morrow and Anthony Weston. Hackett Publishing Company, 2011.

Critical Inquiry: The Process of Argument, by Michael Boylan. Westview Press, 2009.

A classic book about heuristic reasoning in mathematics is

How to Solve It: A New Aspect of Mathematical Method, by George Polya. Princeton University Press, 2004.

Graphic Design

See, e.g.,

Looking Good in Print, by R.C. Parker and P. Berry. Coriolis Group Books, 1998.

The Non-Designer's Design Book, by R. Williams. Pearson Education, 2008.

Engineering Ethics

There are many good books on this subject. See, for example,

Engineering Ethics, by Carl Mitcham and R. Shannon Duval. Prentice Hall, 2000.

Engineering Ethics: An Industrial Perspective, by Gail Baura. Elsevier Science, 2006.

Introduction to Engineering Ethics, by Mike W. Martin and Roland Schinzinger. McGraw-Hill Education, 2010.

Persuasive Speaking

See, for example,

The Challenge of Effective Speaking, by Rudolph Verderber, Kathleen Verderber, and Deanna Sellnow. Cengage Learning, 2011.

Persuasive Business Proposals: Writing to Win More Customers, Clients, and Contracts, by Tom Sant. AMACOM Books, 2012.

Quotations

Some source books for potentially useful quotations include:

Random House Webster's Quotationary, by Leonard Roy Frank. Random House Reference, 2001.

The International Thesaurus of Quotations, by Eugene Ehrlich. Collins Reference, 1996.

The Oxford Dictionary of American Quotations, by Margaret Miner and Hugh Rawson. Oxford University Press, 2005.

The Yale Book of Quotations, edited by Fred R. Shapiro. Yale University Press, 2006.

Designing Slides

slide:ology: The Art and Science of Creating Great Presentations, by Nancy Duarte. O'Reilly Media, 2008.

Presentation Zen: Simple Ideas on Presentation Design and Delivery, by Garr Reynolds. New Riders, 2011.

Slide Rules: Design, Build, and Archive Presentations in the Engineering and Technical Fields, by Traci Nathans-Kelly and Christine G. Nicometo. Wiley–IEEE Press, 2014.

Presentation Patterns: Techniques for Crafting Better Presentations, by Neal Ford, Matthew McCullough, and Nathaniel Schutta. Addison-Wesley, 2012.

Using Powerpoint

PowerPoint 2013 Bible, by Faithe Wempen. Wiley, 2013.

PowerPoint 2013 For Dummies, by Doug Lowe. Wiley, 2013.

Beyond Bullet Points: Using Microsoft PowerPoint to Create Presentations That Inform, Motivate, and Inspire, by Cliff Atkinson. Microsoft Press, 2011.

LaTeX and TeX

Most engineers probably use PowerPoint or one of its relatives (such as LibreOffice Impress®) to make slides. However, an engineer with a strong mathematical orientation may wish to use the LaTeX typesetting system along with a slide-making package such as Beamer. The standard manual on LaTeX was written by the inventor himself:

LaTeX: A Document Preparation System, by Leslie Lamport. Addison-Wesley, 1994.

Also useful is

A Guide to LaTeX: Document Preparation for Beginners and Advanced Users (3rd Edition), by Helmut Kopka and Patrick W. Daly. Addison-Wesley, 1999.

The TeX system on which LaTeX is based was invented by the computer scientist Donald Knuth. His book is

The TeXBook, by Donald E. Knuth. Addison-Wesley, 1984.

Rubrics

How to Create and Use Rubrics for Formative Assessment and Grading, by Susan M. Brookhart. ASCD, 2013.

Introduction to Rubrics: An Assessment Tool to Save Grading Time, Convey Effective Feedback, and Promote Student Learning, by Dannelle D. Stevens and Antonia J. Levi. Stylus Publishing, 2012.

Assessing for Learning: Librarians and Teachers as Partners, by Violet H. Harada and Joan M. Yoshina. Libraries Unlimited, 2010.

Creating and Recognizing Quality Rubrics, by Judith A. Arter and Jan Chappuis. Pearson Education, 2007.

Scoring Rubrics in the Classroom: Using Performance Criteria for Assessing and Improving Student Performance, by Judith A. Arter and Jay McTighe. Corwin Press, 2001.

Appendix: Presentation Checklist

This checklist will take you step-by-step through the complete technical presentation process, from planning, through preparation and practice, to the completion of your talk and what to do afterward.

Engineering Your Presentation

☐ **I understand the goal of my presentation**
☐ I understand how the relevant information resides in my mind
☐ I fully understand the target audience
☐ *I know the background of the audience*
☐ *I know the purpose of the listeners*
☐ *I know the level of understanding of the listeners*
☐ **I have completed the research necessary to generate solutions to my speaking task**
☐ I know the time limit
☐ I know the limitations of the venue
☐ *I know how the audience will be positioned*
☐ *I know what presentation equipment will be used*
☐ *I'm aware of the acoustics and lighting*
☐ **I have generated several potential solutions to my speaking task**
☐ **I have evaluated the potential solutions and decided on the one that best meets my needs**

Designing Your Presentation

☐ **I have established the structure of my presentation**
☐ **I have designed my introduction**
☐ I understand what I want to accomplish with my introduction
☐ I have decided on a structure for my introduction
☐ **I have designed the main body of my presentation**
☐ I know the type of presentation I will be doing
☐ I have decided on an order of presentation
☐ I have considered how I will persuade the listener

☐ *I have a way to gain the listener's attention*
☐ *I address the need for my product*
☐ *I offer a solution*
☐ *I differentiate my product from that of the competitors*
☐ I have created an outline of my talk, possibly using a mind map as a guide
☐ **I have designed the conclusion of my presentation**
☐ I understand what I want to accomplish with my conclusion
☐ I have an indicator of the end of the presentation

Building Your Presentation

☐ **I have established the target specifications for my talk**
☐ I have established the scope of my talk
☐ I know what topics I want to cover
☐ I have thoroughly investigated the motivations and interests of the audience
☐ I understand the time limitations
☐ I understand the venue limitations
☐ **I have prepared my visual aids**
☐ I have selected appropriate presentation software
☐ *I know whether specific software must be used*
☐ *I am familiar with the software or willing to learn*
☐ I have created a slide template
☐ *I know whether a format is required*
☐ *My template includes helpful information*
☐ Title
☐ Page number
☐ Graphical tag (e.g., organizational logo)
☐ Footer or header with date/venue/presenter etc.
☐ *My template is clean and uncluttered*
☐ I have decided on the number of slides to use, based on time limit and audience needs
☐ I have checked the appearance of all my slides
☐ *Fonts are appropriate*
☐ I did not mix too many fonts
☐ The size is sufficient to see from all points in the venue
☐ *The amount of material on each slide is appropriate*
☐ The number of equations, tables, and figures on each slide is appropriate

☐   I did not combine too many elements

☐   *I did not use too much text*

☐   There is not more text than the audience can easily read

☐   I only include enough text to provide myself cues and give the audience a point of focus

☐   *Color is appropriate*

☐   I chose color carefully

☐   I did not mix too many colors

☐   Color combinations are distinctive

☐   I considered the impact of color on audience members who are color blind

☐   *Slide background is appropriate*

☐   The background is not too dark

☐   The background is not too busy

☐   The background is not distracting

☐   *Adjunct elements are appropriate*

☐   Animation is not overbearing or otherwise distracting

☐   Sound or movies add to the impact of the presentation and are not just for show

☐   Active links to web pages do not disrupt the flow of the presentation

☐   *My equations, graphs, and tables are easy to read*

☐   Font size is not too small

☐   Lines are not too thin

☐   Captions are short, easy to read, and meaningful

☐   Legends are clear and easy to understand

☐   There are not too many elements on any slide

☐   *I have checked the appearance of my presentation using the venue equipment or similar equipment*

☐   The presentation is visually appealing

☐   All fonts, equations, and graphs appear as expected

☐   All adjunct elements (animation, links to web pages, sound, movies) behave as expected

☐ **I have carefully examined my materials for any logical blunders**

☐   There are no fallacies of formal logic

☐   There are no fallacies of informal logic

☐ **I have checked my materials for errors in technical claims**

☐   My claims are reasonable

☐   My claims pass dimensional and order-of-magnitude checks

☐   My claims agree with known special cases

☐   My claims are in accord with known standard principles

☐   My evidence comes from reliable sources

☐   I have examined relevant counterexamples to my claims

☐ **I have checked my materials for proper English usage**

☐ My word choice is appropriate

☐ *Words have the proper technical meaning*

☐ *Words are at the proper level of formality*

☐ The punctuation and spelling obey standard rules

☐ My wording takes the form of either description or argumentation, as appropriate

☐ *My descriptive content has a clear standpoint*

☐ *My descriptive content has the appropriate level of detail (no more, no less)*

☐ *My descriptive content forms a coherent picture*

☐ **I have checked my materials for proper mathematical discourse and clarity of presentation**

☐ I have used proper mathematical notation

☐ I have not overburdened the listener with too many equations

☐ My equations form a logical progression

☐ I understand what I want to say about each equation

☐ **I have carefully planned my demo**

☐ The demo is easily seen by all audience members

☐ There are no safety issues

☐ I have a backup plan in case of loss or damage of the demo equipment

☐ **I have considered the special needs of a poster presentation**

☐ I am aware of the formatting requirements

☐ I know the rules of presentation

☐ I have determined the amount of material I can place on the poster

☐ I have created a workable layout

☐ *I have replaced the temporal orientation of an oral presentation with an appropriate spatial orientation*

☐ *I have created an alternative layout that is clear and easily understood by the viewer*

☐ I have created or gathered needed ancillary items (handouts, business cards, brochures, etc.)

☐ **I have considered any ethical implications of my presentation**

☐ I understand what is appropriate ethical conduct within my discipline/profession

☐ I have consulted the ethical codes for my organization

☐ I understand the special needs of my audience

Optimizing Your Presentation

☐ **I have rehearsed thoroughly**

☐ I have rehearsed in front of an audience member or colleague

☐ I have received feedback from knowledgeable listeners

☐ I have adjusted my presentation based on the feedback

☐ **I completely understand the material I will be presenting**

☐ I understand the terminology

☐ I understand the technical nature of my material

☐ I understand what others in my field have done

☐ I understand how my work fits with what others have done

☐ I understand how my work fits within my discipline

☐ **I am mentally ready to deliver a clear presentation**

☐ I know how to pronounce the terms

☐ I am comfortable with the material

☐ I have identified material that requires special care to explain

☐ **I have my timing down**

☐ I know the time limit

☐ I have identified portions of the presentation I stumble over, and have taken extra care with these passages during rehearsal

☐ I have prepared for possible interruptions during my talk

☐ I have an "exit strategy" if I take longer than I prepared for

☐ **I am prepared for the venue**

☐ I have a pointing device

☐ I have adjusted the volume of my speech to accommodate the room

☐ I have the appropriate clothing

☐ My talk is in the proper format for the presentation equipment

☐ My demo is set up properly for the room

☐ **I am prepared to answer questions**

☐ I am knowledgeable of, and comfortable with, my material

☐ I understand the limits of my knowledge and when to say "I don't know"

☐ I have prepared a list of resources that will be helpful to audience members

☐ **I am mentally prepared to control the audience**

☐ I understand how to deal with unruly or insistent audience members

☐ I am willing to limit my interactions with any given audience member

☐ I understand how to "calm" a distracted audience

Delivering Your Presentation

☐ **I have double-checked that I have all my necessary materials**

☐ **I have given myself sufficient time to travel to the venue**

☐ **I have put myself in the proper frame of mind**

    ☐ I have gotten sufficient sleep

    ☐ I have eaten a good meal

    ☐ I have allowed some time before the talk to 'de-stress' and become calm

☐ **I am giving myself feedback during the talk**

    ☐ I am maintaining eye contact

    ☐ I am speaking with enthusiasm and using appropriate gestures

    ☐ I am speaking with appropriate volume

    ☐ I am aware of my pace and the time remaining

    ☐ I am aware of my personal quirks

    ☐ *appearance*

    ☐ *stance and movement of arms or hands*

    ☐ *overly repeated words or phrases (you know, like, hey, moving forward)*

    ☐ *pause fillers (um, uhhhh, throat clearing)*

☐ **I am prepared to adapt**

    ☐ I have an exit strategy

    ☐ I have rehearsed with an audience

☐ **I have a strategy for handling questions**

    ☐ I repeat questions after they are asked

    ☐ I ask for clarification if needed

    ☐ I defer questions that would take too long to answer

    ☐ I admit when I don't know the answer

    ☐ I am prepared to offer suggestions of outside references

After the Talk

☐ **I congratulate myself for all of my hard work**

☐ **I forgive myself for any missteps or blunders**

☐ **I engage with interested audience members**

☐ **I use the experience to improve**

    ☐ I assess whether I did enough to prepare; what worked and what didn't?

    ☐ I assess how well the talk went and how I can use the experience to improve my delivery

    ☐ I assess whether the talk needs to be changed if I deliver it again

☐ **I consider reinventing my talk for another venue or occasion**

☐ **I find a way to archive my talk**

Index